Ground Improvement by Deep Vibratory Methods

SECOND EDITION

Ground Improvement by Deep Vibratory Methods

SECOND EDITION

Klaus Kirsch

Fabian Kirsch

CRC Press
Taylor & Francis Group
Boca Raton London New York

CRC Press is an imprint of the
Taylor & Francis Group, an **informa** business

CRC Press
Taylor & Francis Group
6000 Broken Sound Parkway NW, Suite 300
Boca Raton, FL 33487-2742

First issued in paperback 2018

© 2017 by Taylor and Francis Group, LLC
CRC Press is an imprint of Taylor & Francis Group, an Informa business

No claim to original U.S. Government works

ISBN-13: 978-1-4822-5756-4 (hbk)
ISBN-13: 978-0-367-13902-5 (pbk)

Library of Congress Cataloging-in-Publication Data

Names: Kirsch, Klaus, author. | Kirsch, Fabian, author.
Title: Ground improvement by deep vibratory methods / authors: Klaus Kirsch and Fabian Kirsch.
Description: Second edition. | Boca Raton : Taylor & Francis, CRC Press, 2017. | Includes bibliographical references and index.
Identifiers: LCCN 2016024076 | ISBN 9781482257564 (alk. paper)
Subjects: LCSH: Soil stabilization. | Vibratory compacting. | Foundations.
Classification: LCC TA749 .K57 2017 | DDC 624.1/51363--dc23
LC record available at https://lccn.loc.gov/2016024076

Visit the Taylor & Francis Web site at
http://www.taylorandfrancis.com

and the CRC Press Web site at
http://www.crcpress.com

Contents

Preface to the Second Edition

Ground improvement by deep vibratory methods continues to be the most used in situ dynamic soil improvement method today, representing well over 10% of all special foundation work. Its versatility in improving the soil conditions in difficult site conditions—be it in view of the treatment depth, over water, or with restricted head room—and its economical and ecological advantages in comparison with other foundation solutions are recognized and form the basis for further development.

Our publishers proposed this second edition that encompasses the content of the first edition, dealing with a number of corrections and necessary clarifications. In addition, it provides a new Section 5.2 on the use of depth vibrators in constructing so-called vibro concrete columns. Although this method improves the soil characteristics only marginally, these small diameter in situ concrete piles are increasingly used for moderately loaded large spread foundations.

All chapters were updated with equipment improvements and have also received new additional case histories highlighting new method applications or special features and advantages. The increasing productivity as a result of the progress made with efficient special plant and effective data acquisition systems has led to a considerable growth of the special foundation market.

In Section 4.6, the partial security concept for slope stability and bearing capacity calculations with stone columns are presented in addition to the conventional safety concept.

Chapter 6, also includes comments on the *Carbon Calculator for Foundations*, which was published in 2013 by the European Federation of Foundation Contractors and the Deep Foundations Institute, providing a simple and efficient procedure to calculate CO_2 emissions of various foundation methods. It is hoped that this tool helps promote those methods that are characterized by high sustainability and low greenhouse gas emissions. Unfortunately, experience tells us that the lowest price of a project is still the decisive factor for the contract award, and as long as the CO_2 emissions of construction works are not considered a price worthy item, you should think that the society is prepared to pay the price for a clean environment these considerations remain just an idealistic exercise.

It is nevertheless hoped that this book will help to appreciate not only the economical and technical advantages of ground improvement by deep vibratory methods, but also to recognize their ecological advantages.

This book is written for civil and geotechnical engineers and for the contractors who are engaged in foundation and ground engineering. Students will find this book helpful in their advanced and postgraduate studies.

March 2016

Klaus Kirsch (Retired)
Keller Group plc

Fabian Kirsch
GuD Consult GmbH

Preface and Acknowledgments
to the First Edition

The use of depth vibrators for the improvement of soils that are unsuitable as foundations for structures in their original state dates back to more than 70 years. Over the course of time, the deep vibratory methods have become probably the most used dynamic in situ soil improvement methods today. They have not only experienced a continuous development of plant and equipment to carry it out in practice, but also design methods have been proposed and refined to predict the degree of soil improvement that can be achieved.

Today the importance of deep vibratory soil improvement is unrivalled among modern foundation measures; the demand for in situ deep treatment of soils that introduces, if any, only environmentally harmless materials into the ground continues to increase with the rising level of awareness for the environment. Continued development of plant and process controls has resulted in considerable increase in production rates.

When looking at the design concepts in use today, the reader will realize that experience with the system in different soil conditions still plays an important role. Sand compaction, because of the relative simplicity and convincing economy of the system, and the familiar testing methods for the estimation of settlements have probably inhibited the development of theories that would allow the calculation of the improved properties of granular material based upon the fundamentals of soil dynamics. Recent developments show encouraging attempts to correlate the characteristic motion of the depth vibrator working in the ground with the achieved soil properties. The composite system of vibro stone columns and the surrounding cohesive soil has become a challenging field for engineers to apply finite element analyses particularly to the foundation of more complex structures and single foundations. This computational tool supports the designer in his endeavor to find the most economical solution by optimizing the foundation system.

It was the son who encouraged the father to embark on this book project, after the latter had collected knowledge and experience in the field of ground improvement over a period of almost 35 years of his professional life. This book aims to give an insight into deep vibratory soil improvement methods, both from the contractor's view (Klaus) and from the perspective of the consulting engineer (Fabian), whose dissertation on experimental and numerical investigations of the load-carrying mechanism of vibro stone column groups provided an interesting aspect in this context. We are grateful to our publisher, Taylor & Francis Group, for its support in bringing out this book, which highlights its continued interest in ground improvement methods.

We hope that this book provides the practicing engineer and the design engineer with a comprehensive insight into deep vibratory soil improvement methods. The historical development of these ground improvement methods is presented in Chapter 2, which is based on an earlier publication, Kirsch (1993) "Die Baugrundverbesserung mit Tiefenrüttlern," in Englert and Stocker (eds) *40 Jahre Spezialtiefbau 1953–1993: Technische und rechtliche Entwicklungen* (Werner Verlag, Düsseldorf, Germany). We thank Werner Verlag for its kind permission to use it in this book.

We also hope that we will succeed in demonstrating the astounding versatility of these methods as reflected in a number of international case histories from recent years. The reader will hopefully also realize that vibro compaction and vibro replacement stone columns are presently meeting—and will continue to do so—all the requirements of sustainable construction. We will demonstrate in a separate chapter that these methods require only natural materials for their execution, leaving behind only a minimal carbon footprint.

Finally, we do hope that this book will help stimulate further work and research on the subject and on those problems that are not yet satisfactorily resolved or are still the subject of controversy to enable further plant development and to improve ground engineering.

We thank Keller Group plc for their noble assistance in preparing this book, for funding the preparation of graphs and diagrams but especially for allowing access to the technical archives, and for allowing use of data in preparing most of the case histories. In this context, our foremost thanks go to Justin Atkinson, the chief executive of Keller. With the publication of this book in 2010, we would like to take this opportunity to extend our best wishes for continued development and success to Keller Group plc, which pioneered deep vibratory soil improvement and which celebrates its 150th anniversary this year.

Invaluable support has also been given by our former and present colleagues on technical matters. Our thanks go especially to Dr. Alan Bell who has critically reviewed the manuscript based on his long experience of working in ground engineering. Thanks are also due to George Burke, Jonathan Daramalinggam, Guido Freitag, Johannes Haas, Dieter Heere, Gert Odenbreit, Dr. Vesna Raju, Dr. Thomas Richter, Raja Shahid Saleem, Raja Samiullah, Dr. Stavros Savidis, Dr. Lisheng Shao, Barry Slocombe, Dr. Wolfgang Sondermann, Reiner Wegner, and Dr. Jimmy Wehr. Special thanks go to Christine Trolle for her patience in supporting us to organize the design of graphs for the manuscript.

In preparing case histories of recent applications of vibro compaction and vibro stone column methods, we gratefully acknowledge the support of special contractors in providing valuable data and information: Keller Holding GmbH for Sections 3.6.1 through 3.6.4 and 4.7.1 through 4.7.3, Hayward Baker Inc for Section 3.6.5, Keller Limited for Section 4.7.4, and Bauer Gruppe for Section 4.7.5.

December 2009

Klaus Kirsch
Keller Group plc

Fabian Kirsch
GuD Consult GmbH

Acknowledgments to the Second Edition

We thank our publisher, Taylor & Francis Group, for its encouragement and support to write this second edition, which shows its great interest in ground improvement methods. Thanks are also due to Dr. Wolfgang Sondermann, the chairman of the German Geotechnical Society and director of Keller Group plc (London, UK) for valuable advice and information on the international foundation market.

We also thank Dr. Wilhelm Degen of Betterground (Germany) for information on special plant and equipment as well as on the newly developed quality control and operator guidance system for stone columns.

We are grateful for the support and suggestions given by our former and present colleagues Johannes Haas of Keller Holding and Nikolaus Schneider of GuD (Berlin, Germany) on the use and development of the deep vibratory methods. Information on recent case histories were gratefully provided by Eric Droof and Jim Hussin of Hayward Baker, Hassan Ahmed and Guido Freitag of Keller Holding, and Jonathan Daramalinggam and Dr. Venu Raju of Keller Asia.

Special thanks go to Prof. Stavros Savidis, Technical University Berlin, Germany, who allowed access to Degebo archives, and to Dr. Ralf Glasenapp, who helped tracing their invaluable documents of the first use of the vibro compaction method.

Thanks are also due to Christine Trolle of Keller Holding, Michael Schluchter of GuD and Gunter Raabe of Gr Grafico, who supported us in designing new line drawings for this book.

Authors

Klaus Kirsch was born in 1938 in Chemnitz, Germany and graduated in 1964 as a civil engineer (Dipl.-Ing.) from the Technical University Darmstadt, Germany. Until 1969, he worked with Gruner AG in Basel, Switzerland, and later he joined Keller in Germany as site and project engineer. From 1972 to 1975, he was heading the R&D department and gained comprehensive knowledge in the use and application of ground improvement technologies worldwide, helping to introduce the vibro replacement stone column technology in the United States. He was instrumental in the development of a new generation of depth vibrators and their worldwide use for all types of structures.

From 1976 to 1985, he was responsible for Keller's overseas activities, mainly in the Middle and Far East and Africa. Major contracts included the drilling and grouting works for large dams such as the Tarbela dam in Pakistan and the Mosul dam in Iraq. After the completion of individual ground improvement contracts in Singapore and Malaysia, he also opened up new subsidiaries for Keller in this part of the world.

From 1985 onward, he was in charge of Keller's operations in Continental Europe and Overseas and was a main board director of Keller Group plc until his retirement in 2001. During this period, new subsidiaries were set up in France and Italy and, following the reunification of Germany, Keller expanded into Eastern Europe with new subsidiary companies in the Czech Republic, Poland, and Slovakia.

After his retirement from active service, he was a consultant to the Board of Keller Group plc until 2008, chairing at the same time its Group Technology Committee and the Supervisory Board of Keller in Germany. He is a member of the German Geotechnical Society (DGGT) and author and editor of numerous publications in ground engineering.

Fabian Kirsch was born in 1971 in Frankfurt am Main, Germany, and studied civil engineering at the Technical University, Braunschweig, Germany, graduating as Dipl.-Ing. in 1997 after spending visiting terms at the Indian Institute of Technology, New Delhi, India, and at the University of Glasgow, Scotland, United Kingdom. He worked from 1998 to 2003 as research engineer at the Institute for Ground Engineering and Soil Mechanics at the Technical University, Braunschweig, Germany, from where he also obtained his Dr.-Ing. in 2004 for his work on "Experimental and numerical investigations of the behaviour of vibro replacement stone column groups."

Since 2004, he has been working with GuD Consult GmbH, Berlin, Germany, a major geotechnical consultant in Germany with special expertise in soil dynamics. He is a managing partner of GuD since 2008 and was responsible for a number of projects in Germany and abroad including tunneling works in Berlin, Leipzig, and Cologne; wind park foundations in North Sea and Baltic Sea; and ground improvement and piling projects for power plants in India, Turkey, and the Philippines. He is a member of the DGGT where he works in Working Group 2.1 on Recommendations for Static and Dynamic Pile Tests. He is also a member of the German Association of Consulting Engineers (VBI), the Research Association for Underground Transportation Facilities (STUVA), the German Port Technology Association (HTG), and the Berlin Chamber of Engineers. Since 2012, he is a government-approved checking engineer for geotechnical engineering and is also the author of many publications in soil mechanics and ground engineering.

Chapter 1

An overview of deep soil improvement by vibratory methods

Realization of structures always makes use of the soil on which, in which, or with which they are built. Whenever engineers find that the natural conditions of the soil are inadequate for the envisaged work, they are faced with the following well-known alternatives of

1. Bypassing the unsuitable soil in choosing a deep foundation.
2. Removing the bad soil and replacing it with the appropriate soil.
3. Redesigning the structure for these conditions.
4. Improving these conditions to the necessary extent.

When the Committee on Placement and Improvement of Soils of the ASCE's Geotechnical Engineering Division published its report *Soil Improvement: History, Capabilities, and Outlook* in 1978, it concluded that "it is likely that the importance of the fourth alternative will increase in the future" (p. 1). Indeed, in the decades since "the need for practical, efficient, economical, and environmentally acceptable means for improving unsuitable soils and sites" (p. 3) has increased.

Before selecting the appropriate soil improvement measures, it is necessary to determine the requirements, which follow from the ultimate and serviceability limit state of design. These are

- Increase of density and shear strength with a positive effect on stability problems.
- Reduction of compressibility with a positive effect on deformations.
- Reduction/increase of permeability to reduce water flow and/or to accelerate consolidation.
- Improvement of homogeneity to equalize deformation.

When leaving aside the method of exchanging the soil, ground improvement can be categorized into compaction and reinforcement methods. Table 1.1 shows a classification of the methods of ground improvement that are of practical relevance today.

Table 1.1 Ground improvement methods

Ground improvement methods				
Compaction		*Reinforcement*		
Static methods	*Dynamic methods*	*Displacing effect*	*No displacing effect*	
			Mechanical introduction	*Hydraulic introduction*

• Preloading
• Preloading with consolidation aid
• Compaction grouting
• Influencing groundwater
	• Compaction by vibration:
	  • Using depth vibrators
	  • Using vibratory hammers
	• Impact compaction:
	  • Drop weight
	  • Explosion
	  • Air pulse method
		• Vibro stone columns
		• Vibro concrete columns
		• Sand compaction piles
		• Lime/cement stabilizing columns
			• MIP[a] method
			• CMI[b] method
			• Permeation grouting
			• Freezing
				• Jet grouting

Source: Sondermann, W. and Kirsch, K. Baugrundverbesserung. In *Grundbautaschenbuch, 7. Auflage. Teil 2: Geotechnische Verfahren*. Hrsg.: Witt, K.J., Ernst und Sohn, Berlin, Germany, 2009; Topolnicki, M., In situ soil mixing, in Moseley, M.P. and Kirsch, K. (eds), *Ground Improvement*, Spon Press, London, UK, 2004.

[a]   Mixed-in-place.
[b]   Cut-mix-inject.

This book deals with the important soil improvement methods that utilize the depth vibrator as the essential tool for their execution. Granular soils are compacted making use of the dynamic forces emanating from the depth vibrator when positioned in the ground. The reinforcing effect of stone columns, constructed with modified depth vibrators, in cohesive soils improves their load-carrying and shearing characteristics, and, as a modification of the vibro replacement technique, vibro concrete columns are constructed in a similar process by further modified depth vibrators to construct small diameter concrete piles in borderline soils with characteristics that cannot be improved anymore by the basic vibro techniques.

Since the development of vibro compaction during the 1930s and vibro replacement stone columns in the 1970s, they have become the most frequently used methods of soil improvement worldwide because of their unrivalled versatility and wide range of application.

As we will see, deep vibratory sand compaction is a simple concept, and, therefore, design and quality control of compaction in cohesionless soils have remained almost entirely empirical. The development of predictive design methods based on fundamental soil dynamics was probably inhibited by the simplicity of in-situ penetration testing for settlement and bearing capacity calculations. The study of the effect of resonance developing during compaction in granular soils surrounding the vibrator and improved mechanical and electronic controls of the compaction process have, only relatively recently, opened opportunities for significant advances in this field.

The introduction of coarse backfill during vibro compaction and the resulting formation of a granular or stone column was a logical and almost natural development when unexpectedly cohesive or noncompactable soils were encountered. The composite of the stone column and surrounding soil stimulated theoretical studies on settlements and shear resistance by applying standard soil mechanics principles.

When soils needing improvement of their characteristics contain layers of organic material or are too soft to allow the safe formation of a stone column, with a modified depth vibrator columns can be built in the ground using dry or liquid concrete as backfill material, which, after curing of the cementitious material, act like small diameter concrete piles, frequently with enhanced soil characteristics below the toe of the pile and alongside its shaft.

Each of the systems, vibro compaction, vibro stone columns, or vibro concrete columns, has its characteristics and method of execution, and even machine types are different for the two systems of ground improvement, as are design principles, field testing, and quality control.

When in motion, depth vibrators send out horizontal vibrations, and are all excellent boring machines in loose sandy and soft cohesive soils.

The horizontal motion emanating from the depth vibrator being positioned in the ground is the distinctive characteristic that differentiates this method from all the other methods, which utilize vertical vibrations. These methods, which are in general less effective for the compaction of granular soils and which cannot be used for the improvement of fine-grained cohesive soils, are generally not recognized as true vibro compaction methods and hence are not discussed here in great detail.

Depth vibrators are normally suspended like a pendulum from a standard or special crane for vertical penetration into the ground. The vibrator sinks by the desired depth, sometimes assisted by water or air flushing, additional weight when necessary, or even by downward thrust developed by special cranes with vertical leaders. In granular soils, the surrounding sand is compacted in stages during withdrawal, and in cohesive soils imported backfill is employed to form a stiffening column.

The choice of technique follows from the soil and groundwater conditions given in site investigation reports. As will be explained later, the grain size distribution diagram of the soil to be improved is a valuable tool for this choice. Sand and gravel with negligibly low plasticity and cohesion can be compacted by the vibrations emanating from the depth vibrator (vibro compaction), while with an increasing content of fines vibrations are dampened rendering the method ineffective. Experience shows that the limit of vibro compaction is reached with a silt content of more than approximately 10%. Clay particles at even smaller percentages (1%–2%) cause a similar effect. In these cases, soil characteristics can only be improved by adding granular material during the process of forming compacted stone columns (vibro replacement stone columns).

Following a historic overview of the development of these soil improvement methods, two chapters deal with the improvement of granular soils by the vibro compaction method (Chapter 3) and of fine-grained cohesive soils by the vibro replacement stone column method (Chapter 4). The chapters include detailed descriptions of the equipment used and the specific physical processes controlling the two methods and provide state-of-the-art design principles, including methods to assess and mitigate seismic risk when applying either improvement method. Quality control procedures, the evaluation of suitable soils, and practical method limitations together with recent case histories complete these chapters. Chapter 5 deals with method variations, related processes and, especially, with vibro concrete columns where the stone backfill, as with stone columns, is replaced by concrete.

In Chapters 6 and 7, environmental considerations and contractual implications of these ground improvement methods will be described.

Chapter 2

A history of vibratory deep compaction

An overview of the development of ground improvement by deep vibratory compaction started in Germany between the two world wars. The worldwide economic crisis from 1929 to 1931 had a severe impact on the German construction industry. The absence of profits prevented badly needed investment. Mass unemployment reached to more than 30% and paralyzed economic activity. Farsighted personalities, however, looked into new working and construction methods at a time when the government under Chancellor Heinrich Brüning (1885–1970) tried to mitigate the worst misery by job-generating measures such as the first program to build highways, which was then intensified under the Hitler regime after 1933. In this context, the need to compact large quantities of concrete, sand, and gravel was the motive for numerous engineers to look for better and more efficient methods.

2.1 THE VIBRO FLOTATION METHOD AND FIRST APPLICATIONS BEFORE 1945

The Keller Company was particularly interested in making a dense and strong concrete by employing a completely new method, as proposed by Wilhelm Degen and Sergey Steuermann. Their idea was simple, but the effect was novel and startling (Johann Keller GmbH, 1935).

The basic principle of the new method is shown by two simple model tests (Figure 2.1):

1. A steel box is filled with concrete aggregate that is compacted by a vibrator situated at the surface while simultaneously a cement suspension is pumped with a moderate pressure through a pipe ending in the lower layer; the cement suspension rises slowly and steadily through the aggregate to the surface—an ideal concrete pulp and, after setting, a particularly strong and dense concrete is created this way, at a very economical cement consumption.

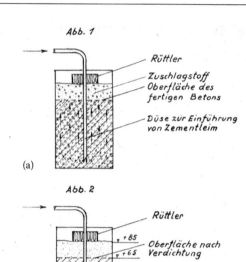

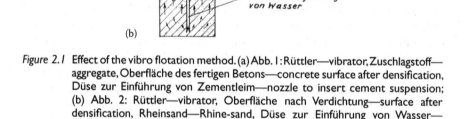

Figure 2.1 Effect of the vibro flotation method. (a) Abb. 1 : Rüttler—vibrator, Zuschlagstoff—aggregate, Oberfläche des fertigen Betons—concrete surface after densification, Düse zur Einführung von Zementleim—nozzle to insert cement suspension; (b) Abb. 2: Rüttler—vibrator, Oberfläche nach Verdichtung—surface after densification, Rheinsand—Rhine-sand, Düse zur Einführung von Wasser—nozzle to insert water. (Courtesy of Keller Group plc, London, UK.)

2. The same box is filled with gravelly sand that is compacted in the same way while water is introduced from below; the sand surface sinks from its original thickness of 85 cm to a height of 65 cm; comparison trials show that no other compaction method achieves an equally strong compression of the sand; even with very intensive vibration, without the rising water the sand thickness is only 72 cm.

This novel method of concrete densification was, for the first and only time, tried in 1934 for the foundation of an I.G. Farbenindustrie building near Frankfurt. The system, which proved so successful even in large scale trials, failed in practice when it was applied to densifying the concrete of a bored pile foundation. The idea of concrete densification was eventually abandoned, and the focus of development of this method was aimed at its possibilities of sand compaction.

The method, shown in Figure 2.1, encompasses all the necessary features that today remain important for effective sand compaction: extensive annulment of internal friction, full saturation of the sand to be compacted, and complete rearrangement of the sand grains with a minimum void ratio as a result of intensive vibrations. Realization of the method required rather complicated watering systems and, above all, powerful surface vibrators to compact the sand in sufficiently thick layers. Even the strongest surface vibrators at the time did not reach deeper than 2.50 m, which limited the new method considerably.

The compaction of sand at large depths was required for the construction of the new Congress Hall at Nuremberg. The designers of this gigantic project had high demands for the durability of its foundations in the light of the aggressive groundwater, and this could not be fulfilled by conventional methods. Also, the dimensions of the structure were exceptional, and so were its loads: the main hall measured 260 × 265 m; the ceiling was 65 m high with a free span of 160 m from wall to wall. The subsoil consisted of sand deposits that extended to 16 m depth where strong sandstone started.

As no proven methods existed for compacting this sand, the designer proposed a field trial to be carried out under the supervision of Degebo (German Research Society for Soil Mechanics). Two compaction methods were to be tested and compared: (1) a modified Franki method and (2) the newly developed so-called Keller method. Trial compaction finally commenced in 1936.

The Keller proposal was as follows:

1. To build vibrators following the principle of a bob weight connected to an electric motor in cylindrical form on a vertical axis with a strong horizontal force
2. To bring the vibrator by means of a bore hole into any desired depth of the sand to be compacted. Figure 2.2 shows the fundamental arrangement of the new method: the vibrator has been lowered inside the borehole to the deepest layer of the sand to be compacted, and the casing has been pulled out to the extent that the vibrator is fully surrounded by the sand. Vibrator and casing are simultaneously extracted in this position with progressing densification. Compaction is accordingly achieved in layers progressing from below to the surface

Loos (1936) reported for the first time before an international audience about this novel method of soil compaction during the first International Conference on Soil Mechanics and Foundation Engineering held in 1936 at Cambridge, Massachusetts in Section M that was particularly devoted to "Methods for improving the physical properties of soils for engineering purposes."

What had been so simply and convincingly formulated at the time, represented in reality a major technical problem: by means of a heavy wooden 23 m high gantry, a 460 mm diameter drilling casing was lowered through 16 m of sand down to the Keuper sandstone employing conventional boring methods. Then the 2 m long vibrator was introduced. It should be emphasized that this was to generate horizontal vibrations in contrast to the

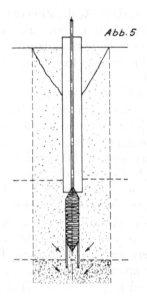

Figure 2.2 Vibro flotation method used at the Congress Hall at Nuremberg. (Courtesy of Keller Group plc, London, UK.)

surface vibrators generally working with vertical oscillations. Compaction was achieved by simultaneously adding water while casing and vibrator was slowly withdrawn at a rate of 8 min./m. The compaction effect was visible at the surface in a crater developing around the casing, into which the excavation material and additional imported sand was backfilled as necessary. Compaction centers were arranged in a triangular grid at a distance of 2.16 m from each other, representing 4.05 m² per probe. Sand consumption was 5.5 m³ for each compaction point, an astounding amount, exceeding all expectations. Degebo confirmed the excellent compaction and raised the allowable bearing capacity from originally 2.5 to 4.5 kg/cm² (250–450 kPa). The unexpectedly favorable results, which were also confirmed by seismic tests, could not obscure grave problems identified during execution. Not only was the method very time consuming, but the vibrator itself, still just a prototype, could only be kept operational at considerable repair expenses. This was the reason that only a relatively minor portion of the foundation work could be performed by the novel vibro flotation method in spite of the convincing test results. The major part of the gigantic foundation totaling some 22,000 compaction centers was performed employing the Franki method compacting crushed gravel and sand backfill (Ahrens, 1941).

Even during the execution of the trial compaction, mechanical improvements were made to the vibrator, extending its operational time. However, it was Rappert, then Keller's project site manager, who had the decisive idea that led to the breakthrough with this method. To complete one compaction point took 22 : 8 h for boring, 12 h for compaction and backfilling, and 2 h to move the gantry to the next position. He proposed avoiding the extraordinarily time-consuming drilling operation by making use of the liquefaction effect of the vibrations on the in-situ sand, enabling the vibrator to sink into the ground under its own weight to the required depth from where the compaction work was to begin by extracting the vibrator slowly back up from the ground. The necessary modifications of the vibrator were made in the same year, so that by 1937 the first foundation works on loose sand with the new vibro flotation method were carried out. Figure 2.3 shows this new method of operation, and Figure 2.4 shows the depth vibrator while executing a trial compaction.

Understandably little was published about the vibrator—the key precondition for the method—and little survived the war in the archives of the Keller Company. After a patent was granted on the method in 1933, the vibrator together with other additional features of the method were also patented in 1936.

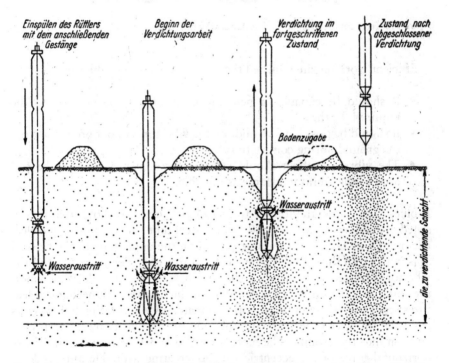

Figure 2.3 Working sequences of the new vibro compaction method. Note: Einspülen des....—flushing down of vibrator and extension tubes, Beginn der....—start of compaction work, Verdichtung....—working stage, Zustand ...—completed compaction, Wasseraustritt—water exit, Bodenzugabe—addition of soil. (Courtesy of Degebo, Berlin, Germany.)

Figure 2.4 Trial compaction at Berlin. (Courtesy of Degebo, Berlin, Germany.)

Efficient work on-site required the following from the vibrator:

- It should be capable of penetrating as quickly as possible to the required depth.
- It should be possible to withdraw the vibrator without problems from the ground during compaction.
- The vibratory effect should efficiently compact a zone in the soil around the vibrator as large as possible.

A blueprint of a depth vibrator from 1937 (Figure 2.5) shows that it consisted of a 260 mm diameter, 2.0 m long, steel tube. In its interior, two electric motors of 12.5 kW each were placed on the upper part of the vertical axis with up to three bob weights at the lower end. The upper part of the vibrator consisted of a device for channeling water to a nozzle at the vibrator point, an inlet for the electric cable for the motors and mechanical features separating the vibratory motions created by the rotating bob weights from the follower tubes carrying the vibrator and enabling its deep penetration into the ground. The motors rotated at 50 Hz resulting in 75 kN horizontal force of the eccentric weights creating a double amplitude of 10 mm. The lifetime of the bearings posed a major problem at the time, but the designers must have quickly succeeded in making the machine robust enough for its use on-site, because in the following years many interesting foundations were carried out with the vibro flotation method.

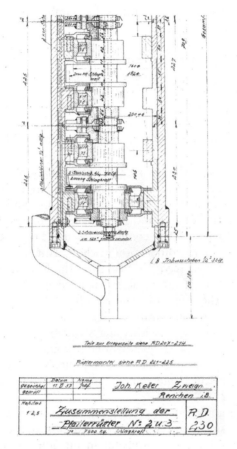

Figure 2.5 Detail from a blueprint of one of the first Keller depth vibrators, 1937. (Courtesy of Keller Group plc, London, UK.)

The new method attracted much interest among experts, although skepticism prevailed in the early years in spite of convincing test results. Leading German ground engineering institutes were involved with trial compactions, and on-sites, introducing various methods to check the densification achieved. Not everybody could be convinced of the compaction results just on the basis of the often surprisingly large amounts of added sand material. Checking and control measures included all known in-situ testing methods at the time, such as direct density measurements, deep soundings, geophysical tests, the exact volumetric measurement of the sand added during compaction, combined with site leveling before and after compaction, and full-scale load testing. As the number of successful applications of the method increased, loose sand, even when under water, lost its poor reputation as unsuitable foundation soil.

In a company catalogue, Johann Keller GmbH (1938) summarized the advantages of sand compaction:

- Naturally deposited sand can be compacted to any depth.
- Artificial sand fill can be compacted over its full height in one operation.
- Densification achieved is complete and permanent; the void ratio can be as low as its theoretical minimum.
- The method is safe, fast, and economic.

These advantages were the reason that increasingly costly pile foundations could be avoided and replaced by shallow foundations on compacted sand with previously unheard-of bearing pressures. Of equal importance were the increased shear strength and the reduced compressibility that accompany the high density of the compacted sand. These findings helped the vibro flotation, or vibro compaction method, as it was also called, to gain an exceptionally broad recognition during the few years before and even during the war. Completed projects included warehouses and industrial plants employing the effect of sand densification at depth to increase the allowable bearing pressure, thus leading to smaller and more economical footing sizes. In connection with water structures at the North and Baltic Seas, the method was also frequently used to compact sand-fill behind walls or dredged fill. Records of some of these projects survived the war and are worth recalling, because interesting features of early vibro flotation works are included.

Degebo was founded in 1928 as a private organization attached to the Technical University Berlin. It accompanied most of these early vibro compaction works and was instrumental in introducing appropriate testing methods for measuring the densification achieved in natural and artificially placed sand deposits with vibro compaction (Muhs, 1949, 1969). In this context, it was found that continuous dynamic penetration tests would not deliver reliable density measurements with increasing depth of investigation because the increasing weight of the steel rod rendered the transformation of the constant energy of the drop hammer increasingly less effective. A static cone penetrometer was therefore developed that was capable of measuring, irrespectively of depth, the point resistance of a cone attached to the rod and being pushed into the ground. Figure 2.6 shows an early measurement of the point resistance by such an electric cone penetrometer to control the density of sand fill before and after vibro compaction.

In 1939, a comprehensive compaction trial was carried out for the foundation of the Great Hall in Berlin. Under the supervision of Degebo, the newly developed method was successfully used in medium sand to a depth of 20 m. Density increased considerably and bearing capacity was more than doubled. To test the capability of the vibrator to penetrate to greater depths, it was suspended from a 40 m high gantry and Degebo (1940)

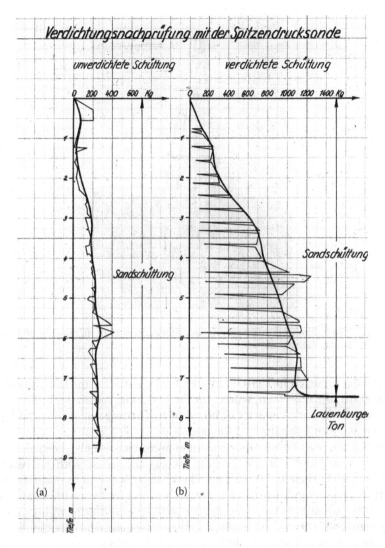

Figure 2.6 Early cone resistance measurement curves in sand fill, (a) uncompacted sand fill, (b) sand fill after vibro compaction. Note: Verdichtungsnachprüfung mit der Spitzendrucksonde—compaction verification with static cone penetrometer, unverdichtete Schüttung—uncompacted fill, verdichtete Schüttung—compacted fill, Sandschüttung—sand fill, and Lauenburger Ton—hard clay. (Courtesy of Degebo, Berlin, Germany.)

observed that the vibrator sank to a depth of 35 m, where it was obviously stopped by the presence of a hard marl layer (Figure 2.7). It was not until 40 years later that modern depth vibrators achieved the same depth again in conjunction with the compaction work carried out for the foundation of the Jebba Dam in Nigeria (Solymar et al., 1984).

Figure 2.7 Vibro compaction employing wooden gantry, 1939. (Courtesy of Degebo, Berlin, Germany.)

Scheidig reported in 1940 on an interesting work for the foundation of a large grain silo at Bremen, where, for the first time, the method was used to increase the bearing capacity of bored piles by compacting the sand below the toe and along the shaft of the piles.

During the war, construction activities were increasingly directed toward military installations or for facilities closely connected with the wartime economy. Were these to be built on alluvial deposits, or in the vicinity of the shoreline, the natural density of the sands was frequently insufficient to carry substantial structural loads using normal shallow foundations without intolerably high deformations. Increasingly in these cases, the vibro flotation method was used by which sand could be compacted sufficiently to carry even the highest loads. In this way, it was possible to avoid expensive pile foundations for which concrete and steel were needed, an effect that was most welcome given the shortage of almost all raw materials. Also, there were major military shelters at Rotterdam and near Bremen where the in-situ or dredged sand was compacted by vibro flotation to such a degree that these heavy structures could be built using incredibly high bearing pressures. At Rotterdam, where a submarine bunker was built, Casagrande proposed an allowable bearing capacity of 15 kg/cm^2 (1.5 MPa) to be used, for which a spacing of the compaction centers of 4 m^2 was necessary. To accelerate the work, the compaction grid was widened to 5 m^2 per probe and the bearing pressure was reduced to 10 kg/cm^2 (1.0 MPa). Figure 2.8 shows a cross section through the submarine shelter.

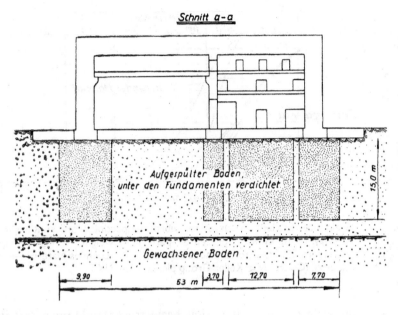

Figure 2.8 Foundations of a submarine shelter, 1942. Note: Schnitt—cross section, Aufgespülter Boden unter den Fundamenten verdichtet—dredged sand compacted below footings, Gewachsener Boden—natural ground. (Courtesy of Keller Group plc, London, UK.)

Of much greater importance in the development of the vibro compaction method were observations made in 1942 in connection with the foundation of an I.G. Farben facility at Rolingheten in Norway. Here fine-grained, loose partially silty to very silty sand was to be compacted at high groundwater table to carry heavy structural loads. The works were supervised by Degebo (1942), which recommended a rather close spacing of 2.5 m² per compaction probe in this case. However, a satisfactory compaction of the sand was only possible through the addition of gravely medium sand as backfill material during execution (Figure 2.9). Backfill material amounted to 20% of the compacted sand volume. Site investigation carried out after compaction revealed for the first time that the core of the compaction consisted primarily of the coarse backfill material, which was compacted to 100% relative density. In the surrounding fine-grained soil between the compaction cores, a relative density of 60% was measured. With today's perception, this compaction work was probably for the first time close to what today is called *vibro replacement*, whereby, as will be shown later, the in-situ fine-grained material cannot be satisfactorily compacted by the vibratory motion alone and coarse backfill material needs to be added. In the Norwegian example, load tests carried out on, and between, the compaction probes revealed an allowable bearing pressure of 6 kg/cm² (600 kPa). The effect of the compaction work was quite impressive: "although even light vehicles could not

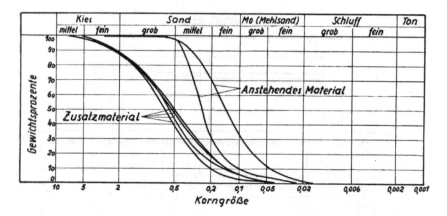

Figure 2.9 Grain size distribution curves for vibro compaction work at Rilingheten, Norway. Note: Gewichtsprozente—percent by weight, Korngröße—grain size in mm, Zusatzmaterial—added material, Anstehendes Material—in-situ material. (Courtesy of Keller Group plc, London, UK.)

pass the site surface before compaction, heavy trucks could run over it after compaction without any difficulties" (Degebo, 1942).

In a relatively short time, the vibro flotation method was acknowledged as an efficient foundation measure and both execution methods and testing procedures were continuously improved. However, even long after the war, slow-moving gantries were still in use that carried the vibrator and could only be moved on tracks. This and the time-consuming method of adding sand using wheelbarrows allowed shift productions of just 50 lin.m of compaction probe or 150–250 m³ of compacted sand.

2.2 VIBRO COMPACTION IN POSTWAR GERMANY DURING RECONSTRUCTION

By the end of the war, the situation of the German construction industry and particularly specialist foundation companies was frightful. Developments that had continued and made progress even during the war came to an abrupt end. Since all efforts were concentrated now on restoring and maintaining the necessities of life, construction activity was restricted to repairing major damage and the restoration of the infrastructure. As the people regained confidence in their future after the currency reform in 1949, and much helped by the Marshall Plan (European Recovery Program) initiated in 1948, construction work on new buildings slowly developed again. The impressive practical results of the vibro flotation method were, however, not forgotten by engineers, consultants, the administration, the industry, and the universities. The method quickly gained increasing importance during the reconstruction phase due to its astounding flexibility and adaptability in solving different construction problems. Whenever the density of naturally deposited or filled sand was insufficient for foundation purposes, the depth vibrator was a useful tool to remedy the situation. Eventually, the outdated gantries were replaced by crawler cranes which, together with other modern equipment, reduced site installation costs and increased production rates considerably.

Besides the densification effect of the depth vibrator as described earlier, its potential to temporarily liquefy fully saturated sand below the water table was soon realized and further developed for practical use. Prefabricated concrete piles and steel tubes as well as sheet piles could be easily lowered into the liquefied zone surrounding the depth vibrator. The process was further supported by strong water flush emanating close to the point of the vibrator. This was of significance as top vibrators and vibratory hammers did not then exist. Many such jobs were successfully executed during the first decades after the war including one spectacular case: a complete lighthouse, Tegeler Plate, was sunk 40 km northwest of Bremerhaven 18 m deep into the sandy bottom of the North Sea by means of six depth vibrators attached to its foundation shaft consisting of a 30 m long, 2.5 m diameter heavy steel pipe.

In 1953, when the planning of a new dry dock at Emden commenced, the designers were considering a special proposal to use the ability of the depth vibrator to sink larger bodies into the ground to install anchors which were necessary for the uplift protection of the dock. Experiments were carried out to establish the optimal size of the anchor blocks together with the number of depth vibrators needed to lower the blocks to the predetermined depth. For the dry dock, 500 cross-shaped concrete anchor blocks were sunk 14 m into the underlying sand layer, simultaneously using two or three vibrators for each block. The 46 mm diameter corrosion-protected steel cables, which were securely fastened in the anchor blocks, extended into the slab of the dry dock from where they were posttensioned with 1 MN. As the sand below the dry dock was also well-compacted, when the vibrators were withdrawn, the system of anchors and dense sand provided an excellent solution for the foundation of the dry dock which was more economical than a conventional gravity structure.

In 1956, at Braunschweig, similar difficulties were experienced during vibro compaction works for an engine shed for the German Railway as at the above-mentioned project in Norway in 1942. Here, the subsoil again consisted of very fine, very silty water saturated sand that was totally liquefied by the vibratory motions, transforming the ground surrounding the vibrator into a viscous liquid. A measurable compaction effect, if at all, was only achieved after a very long time. The vibro flotation method had reached its limit of application. The solution that was finally adopted consisted of lowering the vibrator into the ground without the use of flushing water, withdrawing it completely from the ground after it had reached its final depth and filling the temporarily stable hole left behind by the vibrator with coarse material. By repeated repenetration and backfilling, a stone column consisting of well-compacted coarse material was formed. The in-situ soil was displaced laterally providing, at the same time, a supporting horizontal confining pressure.

To build efficiently such a *vibro replacement stone column*, as the system was soon called, the available depth vibrator proved rather unsuitable. Not only was its surface not smooth enough, and the water pipes attached to it prevented rapid penetration, but the vibrator and its extension tubes were not connected with each other rigidly enough to exercise the desired downward thrust.

2.3 THE *TORPEDO* VIBRATOR AND THE VIBRO REPLACEMENT STONE COLUMN METHOD

The new findings gathered on-sites formed the basis of considerations to redesign the original vibrator which was used unaltered for more than 20 years in sand compaction and to make it suitable for the treatment of cohesive soils. Finally, the Keller Company developed a much improved depth vibrator which was, as a result of its shape, called *Torpedo* vibrator (Figure 2.10). It was characterized by a much more powerful 35 kW electric motor, running

Figure 2.10 The Keller *Torpedo* vibrator. (Courtesy of Keller Group plc, London, UK.)

at 3000 rpm and developing a horizontal force of 156 kN. The double amplitude of the freely suspended vibrator measured 6 mm. Fundamentally new
features were its almost smooth surface and a coupling that was capable of
transmitting substantial vertical forces via the vibrator into the ground stemming from either heavy extension tubes or later from the downward thrust
of the special machines carrying the vibrator. In this way, the vibrator could
achieve penetration into the ground without the assistance of flushing water,
provided the necessary total weight could be applied.

Relatively soon the first foundation works were carried out in only partially saturated silts by forming stone or gravel columns with the vibrator.
These well-compacted stone columns were closely interlocked with the soil
and acted together with the surrounding soil and other adjacent columns
like a normal load-bearing soil. Initially, the stone columns were exclusively
constructed using crane-hung vibrators that were lowered into the partially
saturated soils under their own weight without the help of water. In this
way, the consistency of the soils was not negatively influenced. The vibrator was normally completely withdrawn from the ground and a specified
charge of coarse fill was placed into the cylindrical hole left behind by the
vibrator. Then the vibrator repenetrated and compacted the fill, squeezing
it also laterally into the surrounding soil. By repetition of this sequence, a
strongly compacted stone column was created.

Consequently, by the end of the 1950s, deep vibratory systems were
already being utilized across almost the full range of alluvial soils. Grain
size distribution curves became a prime indicator for the appropriate choice
of ground treatment method (Figure 2.11). When sands and gravel with less
then 10% fines were encountered, vibro compaction was the chosen method
of treatment. In all other cases, the alluvial soils would be improved by the

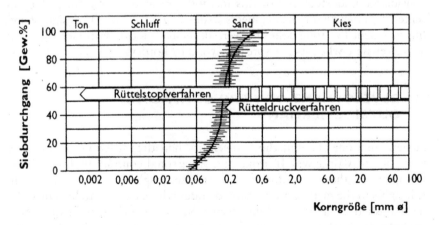

Figure 2.11 Range of application of deep vibratory methods. Note: Siebdurchgang—
percent passing by weight, Korngröße—grain size, Rüttelstopfverdichtung—
vibro stone columns, Rütteldruckverdichtung—vibro flotation. (Courtesy of
Keller Group plc, London, UK.)

vibro replacement method, whereby fine-grained soils with low water content would be treated without the use of water, and those with high water content with the help of flushing water. In both cases, the temporary stability of the cylindrical hole created by the depth vibrator was important and decisive for building a proper stone column of moderate length. The state of the art which the deep vibratory systems had reached in Germany by the end of the 1960s is described by Lackner (1966) in his inaugural lecture at the Technical University of Hannover, underlining the versatility of the method and its economic importance for the treatment of otherwise unsuitable soils.

With the increasing usage of the method, particularly in fine-grained soils, occasional failures and setbacks forced the engineers to think more about the limitations of the stone column method. It was realized that cohesive soils with very high water contents and correspondingly low consistencies could not provide strong lateral support for the stone columns. If the stone columns were placed insufficiently closely packed to each other in these soft soils, which were frequently interspersed by peat layers, overall settlements would exceed expectations. In the absence of accepted design principles for these stone columns which would allow a reliable quantitative settlement prediction, the works on-site were more closely supervised instead. Besides geometrical measurements of the stone columns, the supervision of their steady and consistent build-up by control and registration of the energy used during their construction was introduced. These findings were supplemented by other standard in-situ testing methods and load tests on single columns and column groups. In this way, special contractors and consulting engineers collected valuable information on the behavior of stone columns in different soils, which were very helpful for their further application. The lack of computational methods, however, was increasingly considered a shortcoming of this method, but at the same time it represented a challenge for the engineers. Numerous ideas for the design of soil improvement by stone columns were developed not only in Germany but also elsewhere when its use spread beyond its borders.

2.4 DEVELOPMENT OF VIBRO COMPACTION OUTSIDE GERMANY

Steuermann, one of the method's inventors, had already left Germany a few years before World War II broke out. A personal contract with the Keller Company on the usage of certain joint patent rights granted Steuermann full rights to the method in the United States where he went to live. In 1939, he published an article in *Engineering News Record* on the experiences gained with the method of vibro flotation in compacting sands in Germany. However, it took almost 10 more years before reliable compaction equipment was built in the United States and the first construction works could be carried out.

The new method was called *vibro flotation* in the United States. It only gained full recognition in the early 1950s after the Bureau of Reclamation reported vibro flotation experiments at Enders Dam (1948) and particularly after D'Appolonia (1954) had published his article, "Loose sands—their compaction by vibroflotation."

The vibrator or Vibroflot, as it was called in the United States, was equipped, like its German forerunner, with an electric motor and was operated at 1800 rpm developing 30 Hz with a horizontal force of 100 kN at maximum double amplitude of 19 mm (Figure 2.12). Due to these characteristics, it was ideal for the compaction of sands. The advantages of vibro flotation were seen in the first instance in its easy and controllable working method for effectively compacting loose sands. Soon practical recommendations were developed enabling the engineer to design and supervise the compaction works. The main criterion for judging the success of

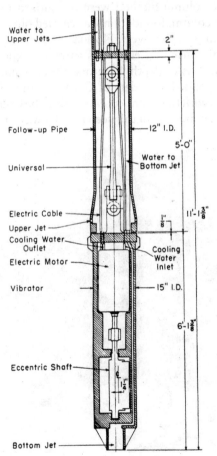

Figure 2.12 Section through Vibroflot. (Courtesy of Vibroflotation Foundation Company brochure.)

the compaction became the relative density of the sand, measured initially directly, but increasingly indirectly using the Standard Penetration Test, and later on also by static Cone Penetration Testing. The general understanding that a safe foundation on sand in a seismic zone required a densification to at least 85% relative density supported the increasing popularity of the method, at first along the east coast of the United States, then in the Midwest, and later also in California. The range of application was from simple foundations for apartment and office buildings to rather complex structures, such as numerous rocket launching pads at the Cape Kennedy Space Centre in Florida (Figure 2.13) and the extensive compaction works at the then largest dry dock at Bremerton in Washington, which the U.S. Navy built in 1961 for the aircraft carriers of its Pacific Fleet (Tate, 1961).

All these works were exclusively for the compaction of sands before the experiences gained in Europe, and particularly in Germany, with the vibro replacement stone column method were recognized in the United States, and the first stone column foundation was carried out in 1972 for the foundation of water tanks at Key West, Florida.

As international consulting engineers gathered experience with ground improvement by means of depth vibrators, the new method of vibro compaction spread beyond the borders of Germany and of the United States. British consultants with their connections within the Commonwealth played a particularly important role in this scenario. By way of license agreements,

Architects & Engineers:
Reynolds, Smith & Hill, Jacksonville, Florida
Maurice H. Connell & Associates, Miami, Florida

General Contractor:
Blount Brothers Construction Company,
Montgomery, Alabama

Figure 2.13 Saturn rocket launching facility at Florida. (Courtesy of Vibroflotation Foundation Company brochure.)

vibro flotation works were already being carried out by the late 1950s in these countries using vibrators from Germany and the United States. From about 1965, the Cementation Company in the United Kingdom built its own vibrators with similar features to the American equipment. However, in 1968, the electric motor was replaced by a hydraulic drive.

More important than these mechanical modifications was the fact that depth vibrators were now increasingly also used in the United Kingdom for the improvement of fine-grained cohesive soils. The application of stone columns in soft soils was accompanied by efforts to develop suitable criteria for the design of this ground improvement method (e.g., Thorburn et al., 1968).

Vibro flotation foundations and, since 1968, also vibro replacement stone column foundations were carried out on some very large international projects, and this made the method known on all continents.

From 1955 to 1958, extensive compaction trials were carried out by the Keller Company to establish the design criteria for the foundation of the Aswan High Dam (Sad al-Ali) now retaining Lake Nasser in Upper Egypt. The effect of vibro flotation performed under water in hydraulically placed fine to medium dune sand was investigated and compared with the results of compaction by blasting. It was found that densification of the sand by vibro compaction was superior to that by blasting (Figure 2.14), and the Board of Consultants for the Sad al-Ali Commission, led by Terzaghi, recommended compacting approximately 3.4 million m^3 of dredged sand through 35 m of water under the core of the dam by this method. The Suez crisis of 1956 and the political tensions thereafter rendered the financing of the project difficult and prevented the contract award to an originally favored German construction consortium. The gigantic project was finally financed and built by the Soviet Union, and purpose-built Russian vibrators were used to compact the sand from 1965 to 1967. Apart from the information that these machines were operated at 1750 rpm, no other characteristics became known, nor did the equipment appear again on the international market.

Of similar scale was the foundation work carried out in 1960–1963 for the Usinor steel works at Dunkirk in France. The subsoil consisted of alluvial fine sands to a depth of up to 30 m, frequently interspersed by silt and clay layers, occasionally even peat. Stiff clay followed below requiring relatively deep embedment of an alternative piled foundation. The economical and practical advantages of a vibro compaction solution convinced the client to follow the recommendations of its foundation consultant, Kerisel, for shallow foundations on vibro-compacted sand, in preference to the piled option. The vibro approach additionally enabled detailed planning to be carried out after the whole area was first compacted, providing program advantages for projects of this size. In less than two years, some 480,000 lin.m of compaction were carried out. Where necessary, the cohesive layers were bridged by large diameter vibro replacement stone columns built by two vibrators (in places even four) suspended 80 cm apart from each other from one crane using particularly strong water flush. The works were performed in double

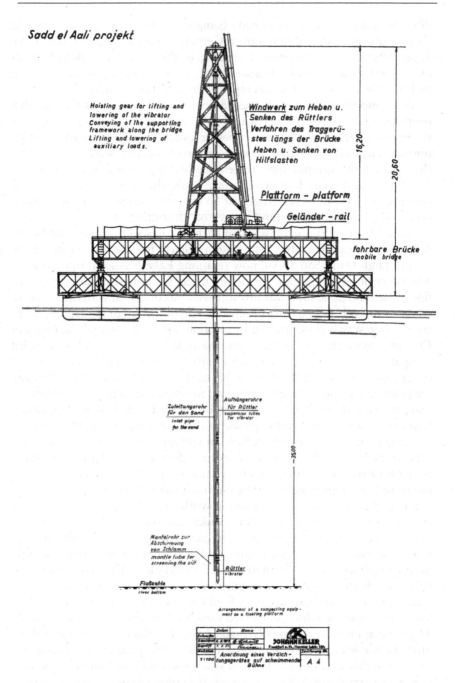

Figure 2.14 Compaction trials equipment for the Aswan High Dam. (Courtesy of Keller Group plc, London, UK.)

shifts employing up to 15 vibrators simultaneously. Compaction results were checked continually by static cone penetration tests. The settlement performance of the structure was as expected by the soil mechanics consultants. Measurements conducted over many years after the plant was put into operation showed only 2–3 cm of settlement even of the heaviest footings.

After India and Pakistan gained their independence in 1947, lengthy and difficult negotiations over the usage of the water from the main tributaries of the Indus flowing from India to Pakistan commenced. The agreement eventually reached was laid down in the Indus Valley Plan, a project financed and controlled by the World Bank and which finally started in 1960. The plan foresaw a series of barrages built on loose sandy river deposits within deep excavations reaching deep below groundwater table. For load transfer purposes and to prevent liquefaction in case of earthquakes, the in-situ sands below the barrages and distribution structures had to be compacted to depths of 20 m. At Sidnai (1963), Mailsi (1965), Marala (1966), Rasul (1966), and Chashma (1968), British and German vibrators compacted more than 2 million m³ of sand. Figure 2.15 gives an impression of the considerable size of these works showing the deep excavation for the Chashma barrage at the time of the vibro compaction works.

When the works in Pakistan were still in progress, quite complex vibro compaction works had to be designed and carried out for an underwater tunnel that was to connect the cities of Parana and Santa Fe in Argentina. The 2400 m-long road tunnel was assembled from prefabricated concrete elements, each 10.8 m in diameter and 65.5 m long, which were placed by a lifting platform into a trench dredged out in the river bed. The double-lane

Figure 2.15 Vibro compaction works in progress at the Chashma barrage in Pakistan. (Courtesy of Keller Group plc, London, UK.)

tunnel at its deepest has its crown 17.7 m below water level and covered by 3.5 m of sand. To create a safe foundation and to minimize the lateral pressure, the in-situ and dredged sand below and around the tunnel was compacted. Some 125,000 lin.m of deep compaction were carried out in 1968 under difficult site conditions.

One example for the advantageous use of the deep vibratory methods in harbor construction is the foundation works for the quay wall of Thuwal harbor on the west coast of Saudi Arabia. About 160,000 lin.m of vibro flotation in loose coral sands were carried out in 1978 for this project from a working ship equipped with five vibrator units. Numerous and often quite extensive foundation works employing deep vibratory methods were carried out on the Arabian peninsula starting in the early 1970s for the infrastructure as it exists today: large power plants and desalination units and impressive facilities for the petrochemical, steel, aluminum, and cement industry. Vibro compaction was and is regularly recommended by the consulting engineers in this area when, besides the cost advantage of the system, the groundwater is particularly aggressive against concrete. This concrete attack, which is especially strong on piles, can be better controlled and indeed avoided by the choice of a shallow foundation placed on improved ground whereby the direct contact with the aggressive groundwater can be avoided or minimized.

Although the first phase of a large grain terminal at Kwinana at the west coast of Australia was founded in 1969 on driven piles, the owner opted for vibro compaction with depth vibrators for the second phase of the project. Situated near Perth, this grain terminal with its loading facilities is among the largest of its kind. The in-situ fine-grained sand was compacted to a depth of 24 m to safely carry the heavy loads of the silos, to reduce the overall settlement of the structures, and to increase their safety against earthquakes. After just 14 months, some 260,000 lin.m of vibro compaction was completed in 1974. The foundation works for this impressive grain terminal won the 1974 Construction Achievement Award of the Australian Federation of Construction Contractors which helped the method of vibro compaction to gain acceptance also on the fifth continent (see Figure 2.16). This was followed by some interesting works at Botany Bay Harbour south of Sidney and then some vibro replacement works for the foundation of railway embankments on soft ground.

In 1962, the Ughelli Power Plant in Nigeria had its foundations laid using the vibro compaction method. In 1972, the method was again used for the foundation of the Massingir Dam in Mozambique. Complex compaction work became necessary for the construction of the Jebba Dam in Nigeria in 1982 and 1983, where deep underlying noncohesive, loose river deposits had to be compacted up to 90% relative density to fulfill earthquake design requirements. The necessary densification to depths of 45 m together with the large volume to be compacted in a relatively short time pushed the limits of the technology and the existing equipment. It was only achieved by a combination of vibro compaction to 30 m with deep blasting for the sand below.

Figure 2.16 Areal view of the Grain Terminal at Kwinana, Western Australia. (Courtesy of Keller Group plc, London, UK.)

By the end of the 1970s, soil improvement using depth vibrators was already an internationally accepted foundation method, which had not only passed its acid test on all continents but enjoyed great popularity thanks to numerous publications on very different fields of application. In 1973, Breth described in a German publication on nuclear reactor safety the importance of vibro compaction to reduce the liquefaction potential of saturated sands during earthquakes. At the same time, interesting studies and field trials on vibro replacement stone columns were carried out in California in conjunction with the foundation of a sewage treatment plant (Engelhardt and Golding, 1973). The investigations were performed to study the behavior of stone columns in a seismic environment. The findings were indeed very encouraging and extended the field of application of the method considerably. It was demonstrated that vibro replacement stone columns provide very effective drainage paths helping to reduce the excess pore water pressures that normally develop in saturated soils during a seismic event. Due to their large shear resistance, the stone columns can also safely carry the horizontal loads that develop during an earthquake as a result of the horizontal ground acceleration. Figure 2.17 shows a picture of a horizontal load test to identify the stone column's shearing resistance. Subsequently numerous foundations employing the vibro compaction and replacement methods were carried out in the seismic zones of Europe, North America, Africa, and Asia and have performed successfully since.

Figure 2.17 Preparation of a horizontal load test on vibro stone columns. (Courtesy of Keller Group plc, London, UK.)

2.5 METHOD IMPROVEMENTS

During the first years after the war, the vibro flotation method continued to use the rather ponderous wooden gantries that traveled on rails and from which the vibrators were suspended and operated by standard rope winches (Figure 2.7). Soon, however, this slow setup was replaced by standard crawler cranes and, for moderate compaction depths, by purpose-built lifting gear, the so-called *vibrocats* (Figure 2.18), which resulted in a considerably better output. With

Figure 2.18 First generation vibrocats, 1956. (Courtesy of Keller Group plc, London, UK.)

the increasing application of the vibro replacement stone column method, it was found that the vibrator would not penetrate fast enough into cohesive soil with relatively low water content. A remedy was found initially by increasing the weight of the extension tubes of the vibrator. But the additional load increment was rather limited. Only after the vibrocats were modified and improved in such a way as to allow a downward thrust to be effected on the vibrator was this handicap eliminated and satisfactory performance also achieved in these soils. This new setup provided a considerably improved process security and enhanced productivity when compared with the crane-hung method, and this was welcomed for technical reasons by supervising engineers and for economical reasons by specialist contractors (Figure 2.19).

The construction of a stone column, using the dry top feed method, requires the vibrator to be completely withdrawn from the ground to allow the stone or gravel to be filled in scoops into the hole created by the vibrator. This modus operandi reaches its limit in very soft, saturated soils when the vibrator hole tends to collapse immediately after the vibrator has been withdrawn from the ground. In these circumstances, vibrator hole stability can be obtained by using water that emanates from the vibrator point, creating an annular space around the vibrator, through which the backfill material can be fed into the hole. The coarse backfill sinks to the bottom of the hole where it is compacted by the vibrator, which is withdrawn in stages from the ground. The flushing water is loaded with soil particles and flows back to the top of the hole from where it needs to be channeled away from

Figure 2.19 Vibrocat developing downward thrust, 1977. (Courtesy of Keller Group plc, London, UK.)

the compaction point generally into a settling pond. Sludge handling and water treatment to allow its safe return into the environment can be time consuming and expensive. However, this wet method is still being used today whenever the local conditions permit.

A major breakthrough was achieved when, in 1972, a patent was registered for the bottom feed vibrator. It allows the stone backfill to be channeled via special piping in the extension tubes, at the outside of the vibrator, and supported by compressed air to be released directly at the vibrator point. The vibrocats were adapted for the new requirements and a special material lock was subsequently developed that allowed the stone backfill to be fed in a controlled way into the extension tubes now acting as material containers. The advantages of the improved method were soon fully realized and were so convincing that the wet method for stone column installation was almost completely replaced by the new bottom feed method. It was not only the sludge-handling problem that was responsible for this development, it was also the new concept itself, which enabled for the first time a safe—and even for critical users visible and verifiable—construction of an uninterrupted stone column. Hitherto occasionally occurring discontinuities or even interruptions in stone columns, which had caused excess settlements, were a key reservation that critics raised against the method. Not only did clients and their consultants welcome the new bottom feed method, but also the practitioners on-sites, because by the new method the quality of stone columns was substantially increased without increasing the accompanying controls.

After the main patent rights to the vibrator elapsed at the end of the 1960s, other specialist contractors, especially in Germany, started designing and constructing their own vibrators. Since hydraulic motors had gained general acceptance in the design of construction equipment, it was not surprising that hydraulic drives were also used for these vibrators. Above all, it was the aim of the designers to increase the stamina and endurance of the vibrator for its rough usage on-sites. Nevertheless, design and development targets for the improvement of vibrator performance were—and are still today—not at all equal among the few companies specialised in the manufacture of vibrators. And it is interesting to see that very little has been published so far about their key characteristics. It is generally accepted today that an optimal ground improvement can only be achieved by an optimal choice of the compaction equipment.

2.6 DESIGN ASPECTS

Although densification of clean sands by vibro flotation is the oldest form of deep vibratory compaction, no computational design method has yet been developed. As the degree of density in given sand achieved by vibro compaction depends on the distance between compaction centers, the vibratory energy consumed and the characteristics of the vibrator itself, it is common practice to determine the necessary probe spacing for the required density by field trials. Correlations of direct and indirect in-situ density measurements with the deformation modulus form the basis of settlement predictions. Seed and Booker (1976) and later on Baez (1995) developed certain design principles for the application of vibro compaction or stone columns to reduce the liquefaction potential of sands in seismic areas.

During the 1970s, the use of vibro replacement stone columns depended very much on the experience gained from numerous projects. As the interest in the method increased with more scientifically orientated engineers, field experience was increasingly complemented by design approaches to predict bearing capacity and deformation behavior of stone columns. Numerous publications on the subject reflect this development during the 1980s. Stone columns improve the ground because they are stiffer than the soil that they replace. Their stiffness depends on the characteristics of both the soil and the stone column material. Although their interaction under load is very complex, reasonably accurate computational methods for predicting settlements do exist for the simple case of the infinite grid of stone columns. Priebe proposed a simple, semi-empirical method in 1976, which he then refined and adapted to better match reality in later years, for the last time in 2003, when he extended his method and formulae into extremely soft soils (Priebe, 1976, 1987, 1988, 1995, 2003). This and other methods have in common the fact that they are not applicable for small groups of stone columns, a case which is of great practical importance. We will see later in Sections 4.3.3, 4.3.5, 4.6.3, and 4.6.4 that this gap is being closed today by numerical methods for the calculation of the bearing capacity and settlement behavior of stone column groups.

On-site, a fully controlled and instrumented construction process of the deep vibratory method is today—80 years after it was for the first time introduced in Germany—as important as is an efficient verification of the success of the soil improvement itself. The flexibility and versatility of the method has widened the spectrum of its use not only geographically but even more so from a technical point of view. It is environmentally completely neutral as it uses only inert and chemically inactive materials, and an overview of its eight decades of history is not just nostalgia but, as will be shown in the following, opens up interesting perspectives for its further development.

Chapter 3

Vibro compaction of granular soils

3.1 THE DEPTH VIBRATOR

The design principles of modern depth vibrators have changed little since the time they were invented and then further developed to suit the needs of practical use on-site. The vibrator is essentially a cylindrical steel tube with external diameters ranging between 300 and almost 500 mm, containing internally as its main feature an eccentric weight at the bottom, mounted on a vertical shaft that is linked to a motor in the body of the machine above. The length of the vibrator is typically between 3 and 4.5 m and its weight ranges from 1500 to about 4500 kg. Figure 3.1 displays these features in a cross-sectional presentation.

When set in motion, the eccentric weight rotates around its vertical axis and causes horizontal vibrations that are needed for the vibro compaction method. The dynamic horizontal forces are thus applied directly to the surrounding soil through the tubular casing of the vibrator, with the machine output remaining constant regardless of the depth of penetration; this is the key factor distinguishing vibro compaction from other methods employing vibratory hammers with their vertical vibrations.

A flexible, vibration-dampening device or coupling connects the vibrator with follower tubes of the same or slightly smaller diameter, providing extension for deep penetration into the ground. These tubes contain water and power lines for the motor, occasionally also air pipes for jets located at the nose of the vibrator and at opposite sides generally just above the coupling.

Motor drive is either electric or hydraulic, powered by a generator or power pack which is generally mounted as a counterweight on the rear of the crane carrying the depth vibrator. Common power ratings of vibrator motors are between 50 and 180 kW, with the largest machines developing up to 360 kW. The rotational speed of the eccentric weight, which can also be split into two or more parts for structural reasons, is with electrically driven machines determined by the frequency of the current and the polarity

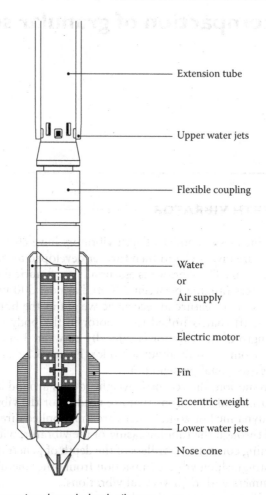

Extension tube

Upper water jets

Flexible coupling

Water
or
Air supply

Electric motor

Fin

Eccentric weight

Lower water jets

Nose cone

Figure 3.1 Cross section through depth vibrator.

of the electric motor. A 50 Hz power source results in a 3000 or 1500 rpm rotational motion with a single or double pole drive. When operating at 60 Hz, the rotational speed is 3600 or 1800 rpm, respectively.

The width of the horizontal oscillation $2a$, or double amplitude, is linearly distributed over the length of the vibrator (see Figure 3.2). It is zero at the vibrator coupling and reaches its maximum—typically between 10 and almost 50 mm—at the point or nose cone of the freely suspended vibrator when operating without any lateral confinement. This is also the point of maximum acceleration $a \cdot \omega^2$ at the vibrator surface, which can reach more than 50 g.

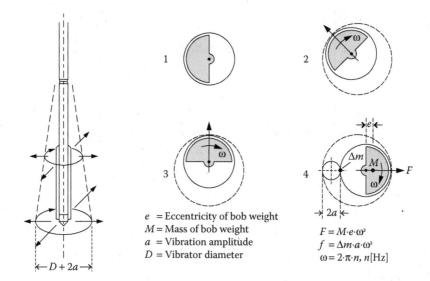

e = Eccentricity of bob weight
M = Mass of bob weight
a = Vibration amplitude
D = Vibrator diameter

$F = M{\cdot}e{\cdot}\omega^2$
$f = \Delta m{\cdot}a{\cdot}\omega^2$
$\omega = 2{\cdot}\pi{\cdot}n,\ n[\text{Hz}]$

Figure 3.2 Principle of vibro compaction and vibrator accelerations in a horizontal plane.

The centrifugal force F results from the rotational speed ω of the eccentric weight with the mass M and an eccentricity of e, and acts as a lateral impact force on the surrounding soil causing its compaction:

$$F = M \cdot e \cdot \omega^2 \qquad (3.1)$$

The centrifugal force F ranges between 150 kN for smaller vibrators and more than 700 kN for the heaviest depth vibrators. When the machine is in normal working conditions, it is restrained by the ground and oscillation amplitudes, and surface accelerations are much less despite the constant centrifugal force F.

Table 3.1 provides an overview of vibrator and motor characteristics of depth vibrators in use today. It is, however, only a selection of a surprisingly large variety of depth vibrators and reflects the ongoing research and development in this field, driven by geotechnical objectives but very often restrained by mechanical and material limits. Hydraulically driven vibrator motors had until recently certain operational advantages because any desired frequency changes are relatively easily effected. However, operational faults may cause unwanted oil spills posing a potential hazard to groundwater and soil unless biologically degradable hydraulic oil is used.

Table 3.1 Selection of currently available depth vibrators (2015)

Specialist contractor	Machine type	Diameter (mm)	Power rating (kW)	Operating frequency (Hz)	Centrifugal force (kN)	Double amplitude (mm)[a]	Special features
Betterground	B 12	292	94	50	170	9	hyd, c
	B 27	354	140	30	270	24	el or hyd, c
	B 41	390	210	30	410	42	el, c
	B 54	460	360	30	842	54	el, c
Keller Group	MB 1670	315	70	50–60	157–226	7	el, v
	LB 20100	315	100	60	201	5	el, c
	S 340/34	421	120	30	340	29	el, c
	S 700	490	290	25–38	742–690	50/15	el, v
	N-Alpha	259	60	60	130	6	hyd, c
Bauer Gruppe	TR 17	298	96	≤53	≤193	12	hyd, v
	TR 75	406	224	≤33	≤313	21	hyd, v
PTC	VL 18	na	113	50	181	na	hyd
Fayat Group	VL 40	na	135	30	145	na	hyd
	VL 40S	na	180	40	258	na	hyd
	VL 110	na	202	28	353	na	hyd

(Continued)

Table 3.1 (Continued) Selection of currently available depth vibrators (2015)

Specialist contractor	Machine type	Diameter (mm)	Power rating (kW)	Operating frequency (Hz)	Centrifugal force (kN)	Double amplitude (mm)[a]	Special features
Balfour Beatty Ground Engineering	HD 130	310	98	50–60	140–202	16	hyd, v
	HD 150	310	130	50–60	200–288	22	hyd, v
	BD 300	310	120	30–36	175–252	28	hyd, v
	BD 400	400	215	30–35	310–426	34	hyd, v
Beijing Vibro	BJ426-75	426	75	30	270	14	el, c
	BJ402S-180	402	180	30	380	22	el, c

Notes:　c = constant operational frequency, v = variable operational frequency, el = electric drive, hyd = hydraulic drive.

[a] At vibrator point.

When the vibrators are in operation, the ground response causes—as well as the dampening of acceleration and oscillation width—a certain reduction in the operational speed of the vibrator motor. This reduction is generally less with electric motors where it reaches up to 5%, corresponding with the slip in asynchronous motors. More recently, electronic control technology by frequency converters has made it possible to operate electrically driven depth vibrators at variable speeds. In order to avoid the otherwise loss of lateral impact force when reducing rotational speed, designers have developed special divisible eccentric weights that increase their eccentricity substantially when reversing the rotational direction to continue compaction at reduced speed.

One of the key objectives of machine design, besides keeping wear and tear low and repair costs within economically acceptable limits, is to maintain substantial amplitude in the ground. This depends on the interaction of vibrator characteristics and soil response—a complex problem especially with soil constraints differing not only from site to site, and with depth on the same site, but also with compaction time at any given depth. It is, therefore, desirable to know as precisely as possible the operational condition of the vibrator at any time and depth during compaction. The specialists working in this field have developed registration and control devices that record a multitude of parameters as a function of time:

- Depth of the vibrator
- Amperage or oil pressure as a measure of the power output
- Operational frequency
- Air or water flow pressure

Although these working parameters are recorded generally and continuously as a standard procedure on every project, temperature and acceleration measurements at selected critical locations inside the depth vibrator are possible and help in protecting and improving the machine characteristics.

Visual display and recording of these working parameters is today essential and allows the operator to perform the compaction work in an orderly, repeatable, and verifiable manner. Temperature measurements at selected locations within the vibrator motor can also be displayed at the control device serving as motor protection in extreme working conditions.

The outer surface of the depth vibrator is subjected to strong abrasive forces from the surrounding soil. They are particularly severe for broken quartz particles but less so when rounded calcareous material is encountered. Aggressive groundwater, and indeed seawater, accelerate corrosion and may require special attention. For these reasons, the vibrator is generally protected by special wearing plates or other protective means on its outside.

The efficiency of the vibratory systems, when working on-site, is very important and all potential problems leading to unwanted interruptions

Figure 3.3 High-voltage cable connector. (Courtesy of Keller Group plc, London, UK.)

of the construction process need to be scrutinized. A special connection of the electric cable leading to the vibrator motor was therefore developed to avoid the otherwise rather time-consuming process of a vibrator change in deep penetrations where the heavy cable has to be drawn through the complete length of the extension tubes. This cable connector (cable joint) is of the same diameter as the depth vibrator, is capable of transmitting an electric current of 600 A at 400 V, is waterproof up to 10 bar, and is situated just above the vibrator coupling (see Figure 3.3).

The instrumentation and control systems in combination with GPS-based setting out of the compaction probes represent only a first step that will eventually lead in future to a fully automated vibro compaction process. The elements of such technology do already exist and, to some extent, are already being used on-site with modern equipment. However, current labor laws and safety regulations are still prohibitive for its practical realization and introduction in most countries.

3.2 VIBRO COMPACTION TREATMENT TECHNIQUE

3.2.1 Compaction mechanism of granular soils

Clean sands in loose to medium dense states are densified as the particles become rearranged more closely. The resulting degree of volume reduction depends on the characteristics of the vibrations, the soil properties, and the duration of the compaction process. In these soils, the practical depth of efficient surface compaction by powerful vibratory rollers is limited to about 1.5 m. When deeper sand deposits need densification, deep compaction methods using depth vibrators are normally used. In contrast to surface compactors and vibratory hammers, depth vibrators agitate the soil by horizontal vibrations.

With increasing radial distance from the vibrator, the ground vibrations are attenuated by the forces acting between the soil particles. Compaction is therefore only possible when their frictional contact is broken by overcoming the residual frictional strength of the soil. Only then will the soil particles rearrange themselves to find a state of lower potential energy, that is, from loose to dense. For this purpose, a minimum dynamic force is required with a dependent acceleration sufficiently high to break the soil strength. Where, despite continuing transmission of attenuated weak vibrations, the resisting forces within the soil prevent further compaction, there is a limit to how far from the vibrator effective vibrations reach.

It has been found that the stability of the structure of granular soils is destroyed by dynamic stresses when a critical acceleration of more than 0.5 g is reached. With increasing accelerations, the shear strength of the sand decreases until it reaches a minimum between 1.5 and 2 g. At this point, the soil is fluidized, and a further increase of acceleration causes dilation. Figure 3.4a shows the idealized relation between shear strength of the soil and the induced acceleration.

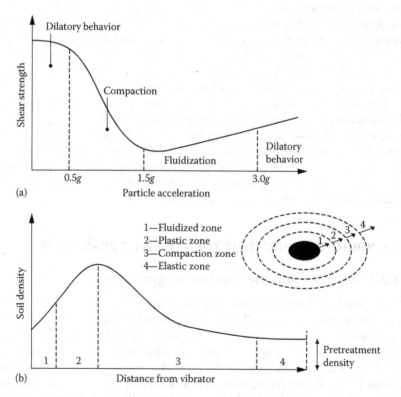

Figure 3.4 (a,b) Idealized response of granular soils to vibration. (After Rodger, A.A., Vibro-compaction of cohesionless soils, Internal Report, R.7/79, Cementation Research Limited, Croydon, UK, 1979.)

In a stage of fluidization, the shear strength of the soil is reduced but not eliminated completely. Therefore, vibrations, although of course dampened, can be transmitted through this zone where particle contacts are continuously broken and remade. As the acceleration transmitted from the vibrator decreases with increasing distance from its source, several annular density zones surrounding the vibrator can be defined as is shown in Figure 3.4b.

In water-bearing soils, fluidization occurs principally when the rate of the pore water pressure increase that is induced by the vibrations exceeds the rate of dissipation, until this pressure overcomes the normal pressure acting between the particles. It may also occur in dry soils by the action of water jetting or when the upward-directed vertical component of acceleration exceeds gravity. As modern machines easily produce accelerations in excess of 10 g, fluidization is induced in the vicinity of the vibrator normally as a combination of the two effects. Soil instability directly caused by the action of acceleration in dry soils is referred to as *fluidization*, whereas in saturated soils the vibrator-induced oscillations cause *liquefaction* depending on pore fluid pressure.

Soil properties that influence the vibro compaction process are

- Initial density.
- Grain size.
- Grain shape and grading.
- Specific particle gravity.
- Depth (influencing intergranular and normal stress level).
- Permeability.

Essential vibrator characteristics influencing compaction are frequency, amplitude, and acceleration of the induced oscillations and out-of-balance force. Finally the duration of the vibration also influences the degree of compaction achieved.

Interpretations suggest that the fluidized zone, which is characterized by minimum shear strength, is a measure of the soil's transmissibility of vibrations and thus is responsible for the radius of influence of the vibratory treatment. In the transitional or plastic zone, the dynamic forces are not sufficient to fluidize the soil but are still strong enough to shear the soil particles from each other at such a rate that they can find a closer packing. From the point of maximal achievable density (Figure 3.4b), attenuation of vibration occurs until it reaches certain threshold shear strength in the ground where any further compaction is inhibited. Water saturation reduces the effective stresses and therefore increases the radius of the compaction zone. It is for this reason that in dry soils the use of flushing water and even flooding of the whole site extends the radius of compaction.

Practical experience gained from construction sites, where depth vibrators with different compaction frequencies were working side by side, has shown that sand can generally be most effectively compacted by vibrating frequencies that are close to what we might call their natural frequency. For this reason, specialist contractors have developed vibrators capable of compacting granular soils using frequencies as low as 25–30 Hz. Occasionally, the accompanying reduction of centrifugal force with frequency was found to be advantageous in optimizing the compaction effect.

Theoretical study of these observations—both in surface and deep compaction—that treat vibro compaction as a *plastic–dynamic problem*, confirm some fundamental findings that have been acquired in practice under operational conditions: for example, at constant impact force the effective range of the vibrations increases with decreasing vibrator frequency, whereas the degree of compaction increases with an increasing impact force. The studies generally aim to develop some kind of on-line control of the vibro compaction by continuously evaluating information obtained from the vibrator movements during compaction. Fellin (2000) suggested making simultaneous measurements of horizontal acceleration in two orthogonal directions at the vibrator tip and coupling. These would, together with the phase angle Φ of the rotating mass, completely describe the vibrator motion when working in the ground. This information could indeed eventually provide the operator with valuable information to better control and direct the compaction work on-site. So far, this additional vibrator instrumentation has only been realized in exceptional cases to support special investigations and scientific research programs.

The study of surface compaction using vibratory rollers also shows an optimal compaction of the soil at its resonance frequency which will generally be between 13 and 27 Hz. These findings compare favorably with the experience obtained from vibro compaction projects indicating that sand and gravel are compacted best by using agitating frequencies of below 30 Hz which are close to their natural frequency.

Model tests carried out in saturated sand also indicate that resonance of the vibrator–soil system could lead to optimal compaction of the surrounding soil. However, in practice, the control of resonance at any point and time during compaction requires sophisticated on-line measurement of vibrator data such as the phase angle Φ between the position of the rotating mass and the vibrator movement and its control during compaction by changing the vibrator frequency f. Figure 3.5 gives the principle of the slip angle measurement, which would form the basis of this control mechanism, with $\Phi = \pi/2$ at resonance. The complexity of such an undertaking in practice becomes evident when remembering the interdependency of vibrator performance characteristics and soil properties with time during compaction.

Today, numerical simulations of the vibro compaction process can also help to better understand this method. It is a well-established fact that granular soils can be better compacted by repeated shearing rather than by compression, and that the degree of compaction achieved depends in the first place on the shear strain amplitude. If the induced strain is too low, the

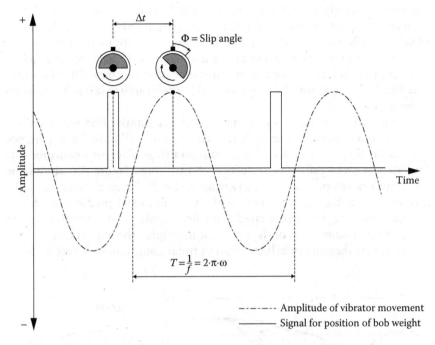

Figure 3.5 Principle of slip angle measurement for resonance control. (After Nendza, M., Untersuchungen zu den Mechanismen der Dynamischen Bodenverdichtung bei Anwendung des Rütteldruckverfahrens, Dissertation, Technische Universität Carolo, Wilhelmina zu Braunschweig, Germany, 2007.)

minimum density cannot be reached even with very large numbers of load cycles; conversely, with large strain amplitudes, optimal densification may not be obtained owing to the effects of dilatancy of the sand.

Based on these principles, a three-dimensional finite element model was recently developed using quartz sand with an initial void ratio of 0.85, a dry density of 1.43 g/cm^3, and a particle density of 2.65 g/cm^3. A disk-shaped element, 0.5 m high and with a radius of 15 m, at a depth of 15 m below surface was modeled. The movement of the vibrators (both a depth vibrator and a top vibrator or vibratory hammer) were simulated using boundary conditions at the surface of a cylindrical hole in the center of the disk. The frequency was chosen as 30 Hz with amplitudes of 7.5 mm, horizontal for the depth vibrator and vertical for the vibratory hammer. By observing the evolution of the void ratio with increasing numbers of strain cycles, three zones were distinguished surrounding both types of vibrators. In the first zone, immediately surrounding the vibrator—radial distance r smaller than 0.5 m—compaction was finished after a few cycles without reaching the minimum void ratio. It is believed that, owing to the relatively large strain amplitudes prevailing within this zone, dilatancy and contractancy balance each other, resulting in a relatively small volumetric change or densification. In the second zone, which extends from 0.5 to 3 m the compaction works best. Strain amplitudes are too small to create dilatancy, allowing the soil to compact. However, with increasing distance from the vibrator, more stress cycles are required to achieve measurable densification before, in zone 3, at a distance of more than 3 m, the strain amplitudes are too small to overcome the intergranular stress level, so no compaction is achieved.

The numerical model not only supports the qualitative findings described above and as shown in Figure 3.4, it also ties in well with the experience gained on-site, particularly that compaction time plays an important role in extending the radius of influence of this zone. Figure 3.6 shows the void ratio development with increasing strain cycles as a function of the distance from the vibrator axis. Although the model predicts too fast a compaction rate when compared with field application, probably due to simplifying assumptions of the model, it highlights the importance of multidirectional shearing in achieving an optimal compaction. While with the

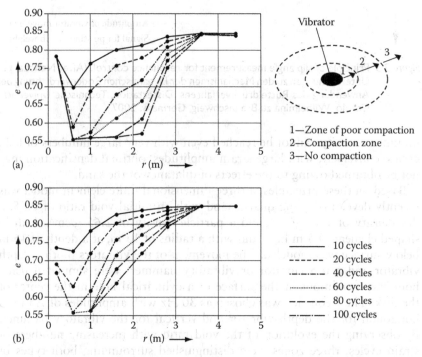

Figure 3.6 Development of void ratio e as a function of the distance r from the source of vibrations (a) for a depth vibrator and (b) for a top vibratory hammer. (Redrawn from Arnold, M. et al., Comparison of vibrocompaction methods by numerical simulations, in Karstanen, M. et al. (eds.), Geotechnics of Soft Soil: Focus on Ground Improvement, 2nd International Workshop on Geotechnics of Soft Soils, University of Strathclyde, Glasgow, UK, 2008.)

depth vibrator, the minimum void ratio in zone 2 extends from 0.5 to 2.2 m after 100 cycles, this zone reaches only 0.5 to 1.0 m for vibratory hammers, simply because the depth vibrator compacts the soil by multiaxial shearing, while the vibratory hammer compaction turns out to be a completely radial symmetric phenomenon resulting in much less multidirectional shearing; this explains the known superiority of depth vibrators in the practice of deep soil compaction.

It is hoped that the findings of newer research work to better understand the complexity of the compaction of granular soil will eventually help to further modernize this technology particularly by obtaining information on the compaction process achieved directly from certain vibrator parameters measured during compaction.

3.2.2 Vibro compaction in practice

With the machine running, penetration of the vibrator and its extension tubes under their own weight is supported by high volume water flow emanating from the lower water jets. At this stage, the jetting water needs to be only at moderate pressure albeit relatively high volume, sufficient to transport loosened sand to the surface through the annulus between the vibrator and the surrounding soil. This material flow allows the vibrator to sink. It is advantageous to maintain a balanced outflow at the surface. Any temporary excess pore water pressure is quickly reduced, again as a result of the high permeability of the granular material in which the compaction work is being carried out. In dry sand, any existing local effective stresses or even slight consolidations from cementation are quickly released and overcome by water saturation and by the shear stresses emanating from the vibrator.

For very deep compactions in excess of 20 m (Figure 3.7), it may be necessary to attach additional water lines with jets at intervals along the extension tubes, or to apply additionally compressed air to maintain or to support the upward water velocity within the annulus against the water losses by seepage from the hole. Should no flushing medium be used in dry sand during the penetration phase, compaction of the material surrounding the vibrator will eventually tighten the soil to such an extent that penetration is stopped. This phenomenon depends—besides the soil properties—mainly on the characteristics of the vibrator. Sinking or penetration rates are generally 1–5 m/min. It is during this penetration stage that the vibrator tends to twist to a certain extent, although fins are generally attached to the vibrator (Figure 3.1) to prevent or at least reduce this behavior, which is the result of the rotational movement of the eccentric mass. Twisted cables and water hoses can be adjusted, if necessary, by removing the vibrator completely out of the ground or, with certain vibrator types, by changing the rotational direction of the electric motor upon reaching the final penetration depth.

When the vibrator attains the design depth, the penetration aids (water and air) are usually switched off or reduced considerably. High pressure,

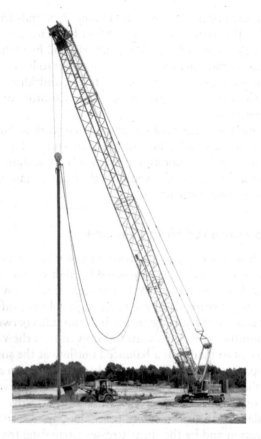

Figure 3.7 Crane-hung depth vibrator, rigged up for 40 m compaction depth. (Courtesy of Keller Group plc, London, UK.)

low volume horizontal water jets situated near the vibrator coupling are now switched on with the aim of undercutting the sand at the sides of the bore and at making it to fall to the vibrator tip, thereby providing a sufficient volume of sand to be compacted. Ideally, the water flow should also be strong enough in terms of volume to maintain a stable bore, balancing any water losses along the bore and at the surface.

Compaction occurs as the vibrator is slowly withdrawn from the ground. It is important at this stage that the vibrator maintains a good and permanent contact with the surrounding soil. When this contact is lost, the vibrator motions may become irregular resulting in insufficient compaction. For better operational control, compaction is normally done in stages of 0.3–1.0 m maintaining the machine steady at each level either for a predetermined time or until monitored vibrator data indicate sufficient compaction having taken place (i.e., power consumption of the electric motor or oil pressure of the hydraulic engine). These time intervals are generally between 30 and 90 s depending on the characteristics of the soil, the required degree of compaction, and the characteristics of the vibrator used.

Whatever the individual operational modes of the specialist contractors may be, it should be remembered that effective compaction requires time and that the unwanted lowering of the machine during the compaction stage may give rise to increased power demand and to a false indication of satisfactory compaction. Densification of the sand surrounding the vibrator causes a reduction of the pore volume, which must be compensated either by extracting it from the material to be compacted, or by introducing imported suitable backfill into the bore from the surface. When no imported sand fill is used and the sand surface is allowed to settle as a result of densification, the surface settlement may reach up to 15% of the compaction depth depending on the original density and the degree of compaction achieved. Independent of the operational mode, an effective regulation of the water flow in the annulus is important. Falsely arranged side jets and badly controlled jetting volumes often lead to the wrong assumption that only by the addition of coarse backfill, which tends to sink more rapidly, can too slow compaction be avoided.

When the cycle of lowering the vibrator to design depth and subsequent stage-by-stage compaction is completed, a new compaction probe can be commenced by repeating the procedure described. The compaction process so carried out leaves behind a compacted sand column of enhanced density, with a dense core followed by material of decreasing density as the distance from the center increases. The diameter of such a column ranges typically between 3 and 6 m depending on the characteristics of vibrator and soil. By arranging the individual compaction points in a grid-like fashion, almost any surface configuration can be treated to the desired depth, both from the dry and offshore.

The distances between compaction points in an equilateral triangular pattern are normally 2.5–5.0 m, and this can cover areas of up to 15 m² for each compaction probe. In this way, up to 20,000 m³ of sand of a favorable grading can be compacted within a 10-hour shift with a single vibrator operation. When twin vibrators can be employed, the production output can almost be doubled, and when special circumstances dictate (for instance large volumes to be compacted offshore with difficult access to site), a multitude of depth vibrators can be used simultaneously safeguarding accuracy of the compaction probe location and the mode of operation (see Figure 3.8).

Typically, for large compaction volumes, the depth vibrators would be suspended from a heavy crawler crane, which also carries an electric generator at its rear. For economic reasons it may be advantageous to operate more than one vibrator from one base machine. In this way, up to six vibrators have been used simultaneously, carried by a specially designed frame for compaction over water (see Figure 3.8b). However, this kind of operation is restricted for practical reasons to a compaction depth of about 25 m.

(a)   (b)

Figure 3.8 (a) Twin and (b) multiple depth vibrators for underwater compaction. (Courtesy of Keller Group plc, London, UK.)

This depth range also represents the maximum depth for the majority of applications. Only in particular circumstances—as were prevailing during the rehabilitation of the derelict open pit brown coal mining areas in Lusatia, Germany—might deeper compaction, in excess of 50 m, become necessary.

As a result of both the vibrator shape and the much-reduced lateral soil pressure close to ground level, densification by the vibro compaction method is generally less effective in the upper 1–2 m below working grade level. If required by the foundation level, this layer therefore needs additional compaction by standard surface compactors before construction work commences.

It is the aim of specialized contractors to continuously better the efficiency of their depth vibrators, not only by improving the compaction characteristics but, equally important, by improving the wear resistance and durability of the equipment, avoiding in this way otherwise expensive overhauls or equipment breakdowns. These generally occur by ordinary wear and tear, particularly in abrasive soils, at the vibrator surface, and internally by excessive heat development in motors and bearings as a result of the compaction work. Selecting appropriate materials and construction parts

Figure 3.9 Fixed mast crane attachment for head room restrictions with telescopic vibrator extensions and cable tightening device. (Courtesy of Keller Group plc, London, UK.)

for vibrators, extension tubes, and attachments is therefore very important. When restrictive working conditions on-site require special solutions, the vibro compaction method often proves its flexibility: in a recent case, the maximal allowed equipment height was restricted to just 10 m, but densification of loose sand was required to a depth of 15 m. The specialist contractor designed and built a telescopic extension for the depth vibrator (and the necessary cable tightening system) attached to the vertical mast of a standard excavator guaranteeing the head room restrictions (see Figure 3.9).

3.3 DESIGN PRINCIPLES

3.3.1 General remarks

The foundation of structures on sand was, for centuries, regarded as a treacherous undertaking, and even today any constructor who intends to build on sand should also beware of sudden settlement of foundations caused by ground vibrations which may be activated by reciprocating man-made and natural impacts from such effects as traffic, vibrating machinery, earthquakes, and even wave action. A sufficient knowledge of the characteristics and behavior of the sand—a clean free-draining granular soil—is a prerequisite of any design attempt aiming to describe its compressibility under load, its shear strength, or its permeability.

Granular soils are characterized by distinctive features that determine their behavior under load. Vibratory deep compaction is mainly concerned with the improvement of the mechanical characteristics of sand. According to the Unified Soil Classification System (USCS), sands contain more than 50% of coarse-grained particles of diameters larger than 0.074 mm, with more than half of the coarse fraction being smaller than 4.75 mm. Clean sand should not contain more than 5% of material which is finer than 0.074 mm. Further subdivisions are made with regard to the nature (silt or clay) of the fine-grained material with grain sizes below 0.074 mm as is shown in Figure 3.10.

Sand is the product of mechanical abrasion of larger stones and boulders during its transport and redeposit. Its geological origin is generally alluvial or glacial, transported by rivers and streams or melt waters flowing from glaciers, sometimes followed by secondary transport process by wind, and finally deposited as sediment. Its most frequent mineral is silica, SiO_2, crystallized as tetrahedron, forming quartz and having a grain density typically of 2.65 g/cm³. It is of some importance whether or not the sand deposit was subjected to glacial loading after its sedimentation, since most mechanical properties of the sand are a function of loading stage or stress history.

The mechanical properties of the sand deposit are also influenced by its grain characteristics such as size, shape, and surface of the particles. Depending on the length of their transportation, sand grains are more or less intensely rounded. In sand, all particles are in contact with each other without any bond or cohesion. Grain packing measured by the relative amount of voids influences the strength and compressibility, the key mechanical properties of the sand. Two different ratios, the porosity n and

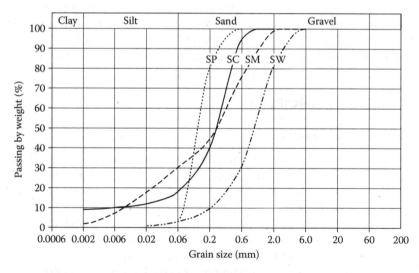

Figure 3.10 Grain size distribution curves of sands (SW, well-graded, clean sand, fines content <5%), SP (poorly graded, clean sand, fines content <5%), SM (silty sand, fines content >12% is mainly silt), and SC (clayey sand, fines content >12% is mainly clay).

the void ratio e, describe the volume of voids, $(V - V_s)$, in relation to the total volume, V, and to the volume of solids, V_s, respectively. Equation 3.2 describes this relationship.

$$n = \frac{(V - V_s)}{V} \qquad e = \frac{(V - V_s)}{V_s}$$

$$n = \frac{e}{(1 + e)} \qquad e = \frac{n}{(1 - n)} \tag{3.2}$$

An idealized sand material consisting of spherical grains of equal diameter can be deposited in two extreme arrangements having a porosity of between $n_{min} = 0.259$ and $n_{max} = 0.476$ or void ratios of $e_{min} = 0.352$ and $e_{max} = 0.924$. These boundaries mark what is called the densest and the loosest packing stage of the deposit. All intermediate stages with porosities n or void ratios e are characterized by their density D or relative density D_r:

$$D = \frac{n_{max} - n}{n_{max} - n_{min}} \qquad D_r = \frac{e_{max} - e}{e_{max} - e_{min}} \tag{3.3}$$

D_r values are slightly larger than D, but describe substantially the same thing (see Table 3.2).

Accordingly the loosest and the densest stages are defined by a relative density of $D_r = 0\%$ and $D_r = 100\%$, respectively. In reality, sand deposits consist of a variety of grain mixtures with boundaries for the loosest and the densest stages falling into the following ranges:

$$n_{max} = 0.39...0.46 \quad e_{max} = 0.64...0.85$$

$$n_{min} = 0.25...0.33 \quad e_{min} = 0.33...0.49$$

Minimum and maximum void ratios of sand are determined by well-defined laboratory tests. The loosest stage of density is achieved by carefully filling a cylinder with dried sand; to achieve the densest stage, the same cylinder is filled with sand while subjecting it to vibrations and static loading. Unavoidable imperfections of the testing procedures, plus the fact that the maximum compaction achievable in the laboratory is sometimes

Table 3.2 Density and relative density of sand

Relative density D_r (%)	Density D (%)	Description
0–15	0–15	Very loose
15–35	15–30	Loose
35–65	30–50	Medium
65–85	50–80	Dense
85–100	80–100	Very dense

below that achieved by vibro compaction in the field, may result in relative densities D_r in excess of 100%.

It is important to understand that all mechanical parameters describing the behavior of sand, such as stiffness and strength or permeability, are directly related to its relative density. The modulus of elasticity of the quartz grain itself, for example, is about 1000 times higher than that of sand, even at its densest stage.

All forces acting on the soil such as gravity or additional loading are transferred by grain contacts. The contact points themselves are able to transfer normal and shear forces, the latter depending on the surface characteristics of the grains. If an idealized sand material is considered to consist of spherical grains of equal diameter, the loosest arrangement is hexahedral whereby each sphere has six contact points with its neighboring spheres. In the densest stage, the spheres form an arrangement which is characterized by 12 contact points with their neighboring spheres, allowing considerably higher stresses per unit volume to be transferred with much less deformations.

As already mentioned, the stiffness of a sand deposit depends on the prevailing stress level and therefore rises generally with depth and the relative density as shown in Figure 3.11.

Beyond a certain threshold, all additional stresses acting on a granular soil are accompanied by a rearrangement of its grains and a nonreversible (plastic) deformation. Unloading of the soil leads to a stress relief and

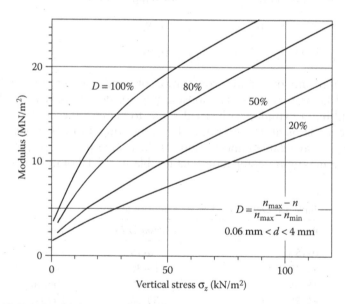

Figure 3.11 Relationship between density D, vertical stress σ_z, and constrained modulus of sand. (After Smoltczyk, H.-U., *Wasser und Boden*, 9, 1966.)

occurs without rearrangement of the grains and with considerably reduced reversed (elastic) deformations. Since this grain rearrangement requires the contact forces between the grains to be overcome, this process can be supported by the lubrication effect of water and by vibrations.

Whenever granular soils are subjected to shear stresses—and practically all loading contains a certain amount of shearing—the grains are forced to rearrange and to find a new stage of equilibrium that is best characterized by their relative density. Depending on the initial density, the applied shear strain can result in compaction or loosening of the sand specimen. This nonlinear deformation behavior of sand is shown in Figure 3.12.

As can be seen from Figure 3.12, dense sand tends to expand under shear stresses, an effect that is called *dilatancy*. We know from Coulomb's theory of friction that the transmissible shear force is a function of the normal force acting on the contact surface. Whenever this volume increases, or dilatancy is restricted in the shear zone—for example, by the adjacent soil—the normal stresses are higher, thereby increasing the necessary shear forces.

Dilatancy of sand has considerable practical importance. It is, for example, responsible for the greater characteristic values of shaft friction to be used in the design for grouted micro piles in sand as compared with the relevant values for bored piles. In general, it is therefore of great economic benefit for almost all geotechnical problems where granular soils are involved to increase the density of natural and man-made sand deposits at least to a medium dense stage.

The design of structures on sand is therefore often concerned with one or a combination of the following features:

- Reduction of settlement by increasing the deformation modulus
- Increase of bearing capacity by increasing the shear strength
- Decrease of liquefaction potential by increasing the density and/or the permeability

Table 3.3 depicts the main characteristics influencing the density of sand that ultimately determine the key properties—modulus, friction angle, and

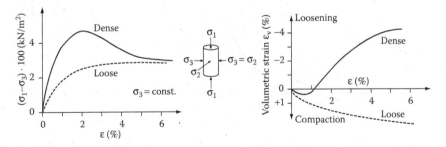

Figure 3.12 Shear stresses in sand (ε = plain strain and ε_v = volumetric strain).

Table 3.3 Physical properties of saturated sand (guideline values)

	Very loose	Loose	Medium dense	Dense	Very dense
Relative density D_r (%)	<15	15–35	35–65	65–85	85–100
N_{SPT} (blows/30 cm)	<4	4–10	10–30	30–50	>50
CPT q_c(MPa)	<5	5–8	8–15	15–20	>20
N_{DPH} (heavy) (blows/10 cm)	<5	5–10	10–15	15–20	>20
Unit weight (wet, above GWT) (MN/m³)	<14	14–16	16–18	18–20	>20
Constrained modulus E_{oed} (MPa)	15–30	30–50	50–80	80–100	>100
Friction angle φ (°)	<30	30–32.5	32.5–35	35–37.5	>37.5
Shear wave velocity V_s (m/s)		<150	220	350	450

permeability—necessary for design. The direct or indirect measurement of the density therefore plays a decisive role in determining the need or otherwise for sand compaction and in any quality control measures.

3.3.2 Stability and settlement control

Natural sand deposits and artificially placed sand fill are likely to have a considerable variation of their key characteristics depending on the nature or method of placement, geological history and, as already explained, the distinctive features of the sand grains (mineralogical origin, size, shape, hardness, and roughness). As we have seen, the potential increase of the fines content inhibits densification, and Figure 3.13 shows

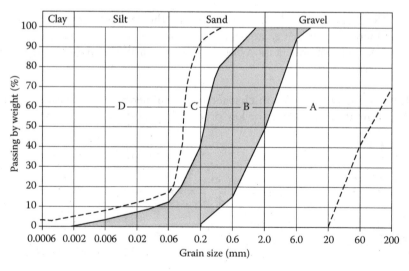

Figure 3.13 Soil grading suitable for vibro compaction. (After Degen, W., Vibroflotation Ground Improvement [unpublished], 1997b.)

the broadly accepted range of grading suitable for vibro compaction as will be further discussed in Section 3.5.

Compressibility, shear strength, and, to some extent, permeability depend primarily on the density of the sand, which is generally described by its relative density D_r in terms of the void ratio or of the dry unit weight of the soil in the loose (L), dense (D), or natural (N) state. Figure 3.14 shows how the settlement of a sand deposit can be assessed based upon the principle of minimum and maximum void ratios.

What appears at first glance to be so simple a procedure turns out to be rather more difficult in practice, because the laboratory work involved in measuring minimum and maximum dry densities of the sand, and then collecting an undisturbed sand sample in the field, and the subsequent determination of its natural dry density, in the laboratory are subject to experimental errors. Not only is this procedure very much dependent upon the experience and accuracy of the soils engineer, it is also very time-consuming and comparatively expensive, particularly in water-bearing soils at greater depths. The direct measurement of relative density in greater depth is therefore today almost completely replaced by indirect methods such as the various available penetration tests that can be approximately related to relative densities.

Table 3.3 gives approximate values for the strength properties of clean, predominantly silica sand, which can be useful for design purposes. However, it must be remembered that sand deposits, natural and artificial, generally vary widely in their properties both in depth and horizontal extension. Thorough site investigations together with precise and noncontradictory specifications for the densification of the sand are therefore indispensable.

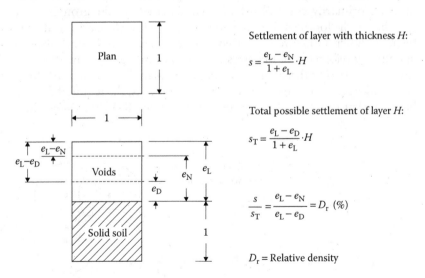

Settlement of layer with thickness H:

$$s = \frac{e_L - e_N}{1 + e_L} \cdot H$$

Total possible settlement of layer H:

$$s_T = \frac{e_L - e_D}{1 + e_L} \cdot H$$

$$\frac{s}{s_T} = \frac{e_L - e_N}{e_L - e_D} = D_r \ (\%)$$

D_r = Relative density

Figure 3.14 Settlement of sand. (After D'Appolonia, E., *Loose Sands: Their Compaction by Vibroflotation*, ASTM No. 156, Special Technical Publication, 1954.)

Experience shows that a compacted free-draining granular soil has a relative density of at least 70% throughout, which is sufficient for foundation
purposes under normal conditions and that any subsequent ground motions
other than earthquakes are generally not strong enough to cause additional
settlement. In the case of seismic events and exceptionally strong impacts
from explosions or similar events, the induced energies can be extremely
high and a detailed study of the required density, which will often need to
be in excess of 85% relative density, becomes necessary (see Section 3.3.3).

As we saw earlier, the effectiveness of the vibro compaction process
decreases with increasing distance from the center of compaction, depending on a number of variables (vibrator characteristics, soil properties, operational modes). Figure 3.4b shows an idealized density profile for a single
vibro compaction point at a selected depth. The decrease from a maximum
in the neighborhood of the center is rather rapid at first, before it lessens
in an exponential way as the distance increases and the original density is
met. By arranging additional compaction points in regular patterns, practical experience shows that the density at the *weakest* point within any
chosen pattern increases broadly by the density increases of the adjacent
probes for single probe behavior at that distance. As it is common practice
today that the specified minimum density for a project is to be met also at
the weakest point of the pattern, this procedure generally provides an additional safety margin because, by definition, all other areas away from this
point are then characterized by higher densities (see Figure 3.15).

The prime objective of vibro compaction treatment is to provide a zone
of improved ground sufficiently dense beneath surface spread foundations.
The settlements of spread footings such as individual pads, strips, or smaller
rafts are primarily controlled by the compactness of the ground immediately below the footings. Compaction points are therefore arranged beneath
them as shown in Figure 3.16. Depth of treatment is very much dependent
upon the development of the pressure bulb, according to elastic theory, generally about two or three times the minimum plan dimension of the footing.
For smaller rafts and closely spaced smaller footings, the required depth may
be somewhat deeper and may often need to reach 8–10 m. Determination

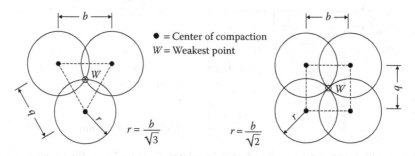

Figure 3.15 Triangular and square compaction patterns.

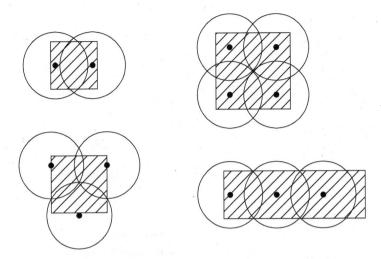

Figure 3.16 Typical arrangements of compaction probes below isolated and strip footings.

of the vibro compaction probe spacing is either by experience from similar cases or by field trials. In view of today's diversity of available depth vibrators, design charts, as have been proposed and widely used in the past (D'Appolonia, 1954; Thorburn, 1975), are not to be recommended anymore. With modern machines, probe spacing can be increased to more than 3.0 m for clean sand, allowing safe bearing pressures in excess of 500 kN/m² with settlements less than 25 mm (see also Greenwood and Kirsch, 1984). One should always remember though that the main objective of settlement control is in most cases to reduce differential settlement. This aim normally does not require compaction to maximum density. Very often, equalization by a moderate widely spaced compaction may be sufficient whereby the sand deposit receives an equally distributed density and stress history for its loaded areas.

Considerably greater depths may become necessary for large raft foundations as well as for dams, embankments or tanks, or where compaction is required to prevent liquefaction or settlement of deeper deposits due to earthquakes or other artificial vibrations. In these cases it is necessary to ensure that the stress levels induced into the soil by vibro compaction exceed those which might occur under design conditions.

We have seen earlier that the radius of influence of vibro compaction depends, for given sand, primarily on the characteristics of the depth vibrator and the chosen mode of operation. It generally varies between 2 and 4 m and it is therefore impracticable and not advisable to specify the necessary spacing of the probes together with the required density criterion. Since the latter is determined by the requirements of the overall design of the project, the necessary spacing follows the choice of machine, and is generally established

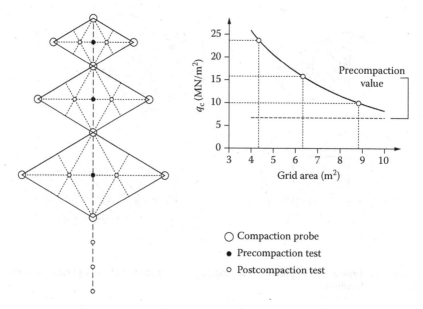

Figure 3.17 Typical arrangement of a vibro compaction field trial to establish probe spacing with exemplary result for specific soil condition and vibratory parameters. (After Moseley, M.P. and Priebe, H.J., Vibro techniques, in Moseley, M.P. (ed.), *Ground Improvement*, Blackie Academic & Professional, Glasgow, Scotland, 1993.)

by experience and, for larger projects, by field trials. Figure 3.17 shows a typical test arrangement of such a trial. If the chosen method of testing is the cone penetration test (CPT) the comparison of pre- and post-CPT results with the desired or specified criterion then determines the necessary compaction probe spacing (see also case history in Section 3.6.6). Field observations show that modern machines with a grid spacing of more than 10 m² per probe in clean compactable medium sand can achieve 80% relative density.

Numerous researchers have studied the behavior of sand and have proposed useful correlations of the various dynamic and static deep penetration tests, in particular the standard penetration test (SPT) and the CPT, with relative density, compressibility, and friction angle. However, the use of these tables and charts (Figures 3.18 and 3.19, Table 3.3) requires a careful validation and interpretation of their applicability in any specific case.

Figure 3.18 provides the possibility of estimating the peak friction angle φ' as a function of the relative density D_r and the gradation characteristics of normally consolidated, saturated, predominantly silica sands for a given stress level in the range of 150 kPa. A similar relationship is shown in Figure 3.19 for uncemented moderately incompressible, predominantly silica sands, indicating that the cone resistance increases linearly with the vertical effective stress for constant friction angles.

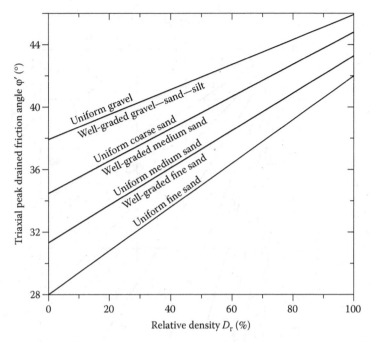

Figure 3.18 Relationship between friction angle φ′ and relative density D_r. (Redrawn from Lunne, T. et al., *Cone Penetration Testing in Geotechnical Practice*, Blackie Academic & Professional, Glasgow, Scotland, 1997; after Schmertmann, J.H., Guidelines for cone penetration test, performance and design, Report FHWA-TS-78-209, 145, US Federal Highway Administration, Washington, DC, 1978.)

For foundations on sand with a smallest width of more than 1.5 m, settlement is normally the key criterion for the design rather than bearing capacity. A reliable in-situ determination of the sand stiffness is of great importance because of the difficulties of measuring the deformation modulus in the laboratory on undisturbed samples obtained from the field as they are extremely difficult to retrieve from granular soils. However, to determine stiffness data for sand from in-situ penetration tests is also rather complex because of their dependence on stress levels, stress history, drainage, and sand characteristics.

For normally consolidated, unaged and uncemented predominantly silica sands, the constrained modulus M can be obtained from the following relations based on CPT results after Lunne and Christophersen (1983):

$$M = 4 \cdot q_c \qquad \text{for } q_c \text{ below } 10\,\text{MPa}$$

$$M = 2 \cdot q_c + 20\,(\text{MPa}) \quad \text{for } q_c \text{ between } 10 \text{ and } 50\,\text{MPa} \qquad (3.4)$$

$$M = 120\,\text{MPa} \qquad \text{for } q_c \text{ above } 50\,\text{MPa}$$

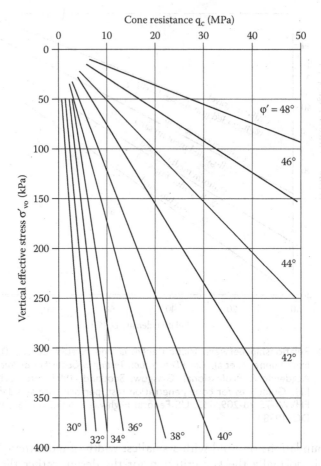

Figure 3.19 Relationship between σ'_{vo}, q_c, and φ'. (Redrawn from Lunne, T. et al., *Cone Penetration Testing in Geotechnical Practice*, Blackie Academic & Professional, Glasgow, Scotland, 1997; after Robertson, P.K. and Campanella, R.G., *Can. Geotech. J.*, 20(4), 718, 1983.)

For overconsolidated sand, the same researchers recommend as a rough guideline:

$$M = 5 \cdot q_c \qquad \text{for } q_c \text{ below 50 MPa}$$
$$M = 250 \text{ MPa} \quad \text{for } q_c \text{ below 50 MPa}$$

(3.5)

When SPT results are considered or specified in a project, Table 3.4 provides a useful guideline for the translation of N-values into equivalent q_c values of the CPT (see also Figure 3.28).

Table 3.4 Proposed translation of N_{SPT} into q_c values for design purposes independent of depth, relative density, and water conditions

Soil type	q_c (kg/cm^2)/N_{SPT} (blows/30 cm)
Silt, sandy silt, and slightly cohesive silt–sand mixtures	2
Clean, fine to medium sands and slightly silty sands	3.5
Coarse sands and sands with little gravel	5
Sandy gravels and gravel	6

Source: Schmertmann, J.H., *J. SMFD*, ASCE, 96(3), 1011, 1970.

As we have seen, correlations between in-situ measured penetration resistances and relative density provide well-established means for the design where silica sands are concerned. However, special care is necessary where sands with high carbonate content are concerned. Calcareous soils with carbonate contents below 50%–70% generally behave similar to noncalcareous soils. Where higher carbonate contents are encountered, these soils often contain up to 100% $CaCO_3$, and the engineering properties are much more difficult to assess. Their coarse grains all derive from shell particles which are subject to grain fracturing during cone penetration and, in addition, very often high cone penetration resistances above the water table are the result of particle cementation due to precipitation, a phenomenon which predominantly occurs in arid countries.

It has been observed that an increasing shell content at first leads to an increase of the cone penetration resistance q_c until grain fracturing becomes a dominant mechanism and the q_c values start to decrease again. Grain fracturing during testing will increase in these sands with increasing shell content, effective overburden pressure, and density. For these sands, measured q_c values tend to grossly underestimate relative density when compared with the results in silica sands which can be up to three times higher at the same density (University Karlsruhe, 2006; Wehr, 2007). When conditions prevail as described above, the performance of load tests or direct density measurements and their evaluation and correlation with CPT values are more appropriate and therefore recommended (see also Section 3.5).

We will see later, in Section 3.3.3, that densification of saturated loose sand represents the key improvement measure to prevent liquefaction during an earthquake, although it is accompanied by a certain decrease in permeability. Consequently, this is apparently counterproductive; however, with density playing the prime role in controlling liquefaction, the method is very effective for this purpose. Conversely in saturated fine-grained sand and silty sands in which densification is difficult to achieve, if at all possible, liquefaction can most effectively be controlled by increasing the overall permeability of the soil mass. To this end, permeable granular vibro replacement stone columns can be installed which shorten the drainage paths considerably, leading to an accelerated pore water pressure relief (see Section 4.3.4).

The above-mentioned effect of permeability reduction by vibro compaction is occasionally used for groundwater control by reducing seepage through existing flood protection dykes or for reducing the ingress of groundwater into deep excavation pits. Sidak (2000) reports on interesting case histories that indicate that vibro compaction reduced the permeability of sand by up to two orders of magnitude, an effect that can often be achieved only by less environmentally acceptable methods.

In addition to the usual application of decreasing the void ratio of granular soils to allow higher bearing pressures and reducing settlement, vibro compaction methods are well-suited to resolving a large variety of geotechnical problems. For the design of retaining structures, vibro compaction can be used to reduce the lateral earth pressure acting on them. The method can also be used below and around piles and caissons to increase their load-bearing capacity, and as the method can also be effectively executed offshore, it is particularly important in construction of harbors and similar work (Kirsch, 1985, and Section 3.6).

As we have seen, density represents the key characteristic of sand, which needs to be measured in-situ to establish most of the other parameters necessary for the design. The flow chart in Figure 3.20 provides soil-inherent (internal) characteristics influencing the density of sand together with external measures by which they can be influenced.

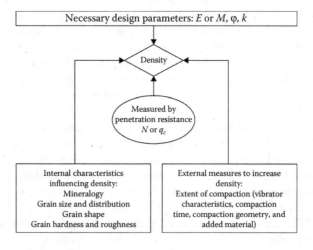

Figure 3.20 Interdependences for the design of structures on sand.

3.3.3 Mitigation of seismic risks

3.3.3.1 Evaluation of the liquefaction potential

It is well-known that saturated medium-to-fine-grained sand may lose its strength during an earthquake. The phenomenon is called *soil lique-faction* and occurs primarily in saturated, cohesionless, fine-to-medium grained soils. In an attempt to explain liquefaction of sand, Casagrande (1936) used the concept of the critical void ratio—dense sand tends to dilate under shear, whereas loose sand undergoes a volume decrease under the same loading condition. The density at which no volume change occurs in the sand under shear load is called the *critical density* (critical void ratio).

Sands having a density below the critical density will therefore settle when subjected to earthquake motions. If drainage is prohibited, the pore water pressure u will increase until it equals the overburden pressure and the resulting effective stress σ' becomes zero. At this point, the sand has lost its strength completely and has become a liquid.

$$\sigma' = \sigma - u \tag{3.6}$$

where:
    σ' is the effective stress
    σ is the total stress
    u is the pore water pressure

The disastrous earthquakes at Niigata, Japan, and in Alaska, both in 1964, triggered studies and investigations on liquefaction caused by earthquakes to better understand the principles and parameters controlling this phenomenon.

Today, the state of the art is best described in the publications of the National Center for Earthquake Engineering Research (NCEER) at the State University of New York.

The *simplified procedure* to assess the liquefaction potential of soils was first developed by Seed and Idriss (1971) and was subsequently periodically corrected and updated with new developments and findings. It deals with level or gently sloping sites over Holocene alluvial or fluvial sediments at relatively shallow depths not exceeding 15 m. The procedure defines the cyclic stress ratio (CSR) that would result from a design seismic event and compares it with the cyclic resistance ratio (CRR), also called *liquefaction resistance*, which the soil is able to mobilize during the earthquake.

CSR is defined in the following equation:

$$\text{CSR} = \frac{\tau_{av}}{\sigma'_{v0}} = 0.65 \cdot \left(\frac{a_{max}}{g} \right) \cdot \left(\frac{\sigma_{v0}}{\sigma'_{v0}} \right) \cdot r_d \tag{3.7}$$

with a_{max} representing the peak horizontal acceleration at ground level resulting from the design earthquake, g the gravity acceleration, σ_{v0} and σ'_{v0} the total and effective vertical overburden pressure, respectively, and r_d the stress reduction coefficient. CSR represents the ratio of the average horizontal shear stress τ_{av} developed by the design earthquake to the initial vertical effective stress before the cyclic loading occurred. Figure 3.21 shows the r_d versus depth curves that are recommended for noncritical projects. The spread of the curves indicates the uncertainty of the method, particularly at depths greater than 15 m.

The simplified procedure replaces the unevenly distributed cyclic shear stresses of an earthquake by an equivalent average uniform shear stress that is equal to 65% of the maximum cyclic shear stress.

Equation 3.7 allows us to calculate the cyclic stresses at different depths and to determine the respective cyclic stress ratios. The magnitude of an earthquake determines the duration of the ground shaking and thus the significant number of stress cycles N_c necessary to generate maximum shear stresses. Table 3.5 provides representative numbers.

By comparing the shear stresses induced by the design earthquake (determining the CSR value) with those shear stresses that are necessary to cause liquefaction under the prevailing site conditions, zones in the soil profile can be identified where liquefaction is likely to occur.

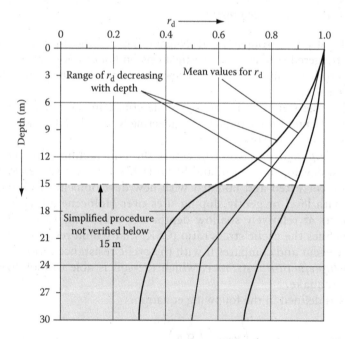

Figure 3.21 Stress reduction coefficient r_d as a function of depth. (After Seed, H.B. and Idriss, I.M., *J. SMFD*, ASCE, 97(SM9), 1249, 1971.)

Table 3.5 Earthquake magnitudes and number of significant stress cycles

Earthquake magnitude M	Number of significant stress cycles N_c
7	10
7.5	20
8	30

Source: After Seed, H.B. and Idriss, I.M., *J. SMFD*, ASCE, 97(SM9), 1249, 1971.

Therefore, the next step is to evaluate the liquefaction resistance of the soil or CRR taking into account the actual soil characteristics that are responsible for this phenomenon.

The difficulties and associated costs for retrieving undisturbed samples from water-bearing granular soil are prohibitive for most projects. Only in exceptional cases and with specialized sampling techniques can sufficiently undisturbed specimens be collected; these are subsequently tested in the laboratory, where the seismic loading conditions can be modeled adequately. Instead, field tests have replaced this procedure and are widely used for routine liquefaction investigations. NCEER recommends four methods: (1) the SPT, (2) the CPT, (3) shear wave velocity measurements, and (4) the Becker penetration test. Since, for the latter two, only limited or sparse test results from liquefaction sites are available, the following deals only with the CRR value measured by SPT and CPT methods.

Figure 3.22 depicts the CRR values as a function of the corrected blow count of the SPT $(N_1)_{60}$ developed by Seed et al. (1985) from empirical data as recommended by the NCEER for 5%, 15%, and 35% fines in the sand. However, these are only valid for magnitude 7.5 earthquakes. The blow count $(N_1)_{60}$ is the measured blow count N_m, corrected to account for the influencing factors, overburden pressure (C_N), energy ratio (C_E), borehole diameter (C_B), rod length (C_R), and sampling method (C_S) by Equation 3.8:

$$(N_1)_{60} = N_m \cdot C_N \cdot C_E \cdot C_B \cdot C_R \cdot C_S \tag{3.8}$$

Suggested ranges of correction factors can be taken from Table 3.6. To account for the overburden pressure, the C_N factor is calculated by using the effective overburden pressure σ'_{v0} in Equation 3.9 that acted at the time that the SPT test was carried out,

$$C_N = \left(\frac{p_a}{\sigma'_{v0}} \right)^{0.5} \tag{3.9}$$

with p_a representing the atmospheric pressure of 100 kPa.

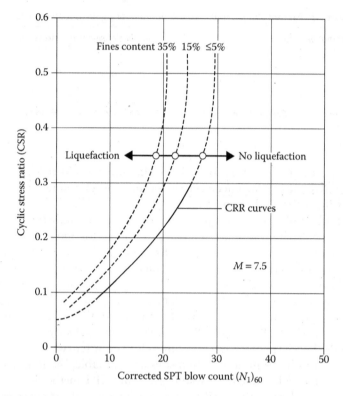

Figure 3.22 Recommended base curve for calculation of CRR from SPT data. (Redrawn from NCEER, in Youd, T.L. and Idriss, I.M. (eds.), *Proceedings of the NCEER Workshop on Evaluation of Liquefaction Resistance*, Technical Report NCEER-97-0022, 1997.)

The SPT-based method of computing the CRR has the advantage that the number of test measurements at liquefaction sites is abundant and that disturbed soil samples can easily be collected during testing to establish fines content and other grain characteristics.

Figure 3.23 shows a similar graph that was developed from empirical data but was based on the relationship between the corrected CPT tip resistance $q_{c,1N}$ and the CSR, showing CRR curves for clean sands with fines contents below 5% and valid for a 7.5 magnitude earthquake. Normalization of the measured cone penetration resistance q_c is achieved by correcting it for the effective overburden pressure σ'_{v0} by

$$q_{c,1N} = C_Q \cdot \left(\frac{q_c}{p_a} \right) \tag{3.10}$$

with $C_Q = (p_a/\sigma'_{v0})^n$ and $p_a = 100$ kPa, the atmospheric pressure, and n ranging from 0.5 for clean sands to 1 for clays.

Table 3.6 Corrections to SPT values

Factor	Equipment variable	Term	Correction
Overburden pressure		C_N	$C_N = (p_a/\sigma'_{v0})^{0.5}$
			$C_N \le 2$
Energy ratio	Donut hammer	C_E	0.5–1.0
	Safety hammer		0.7–1.2
	Automatic-trip donut type hammer		0.8–1.3
Borehole diameter (mm)	65–115	C_B	1.0
	150		1.05
	200		1.15
Rod length (m)	3–4	C_R	0.75
	4–6		0.85
	6–10		0.95
	10–30		1.0
	>30		>1
Sampling method	Standard sampler	C_S	1.0
	Sampler without liners		1.1–1.3

Source: NCEER, in Youd, T.L. and Idriss, I.M. (eds), *Proceedings of the NCEER Workshop on Evaluation of Liquefaction Resistance.* Technical Report NCEER-97-0022, 1997.

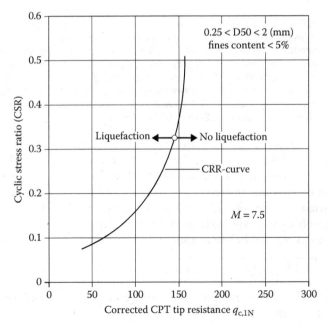

Figure 3.23 Curve for calculation of CRR from CPT data. (Redrawn from NCEER, in Youd, T.L. and Idriss, I.M. (eds.), *Proceedings of the NCEER Workshop on Evaluation of Liquefaction Resistance,* Technical Report NCEER-97-0022, 1997.)

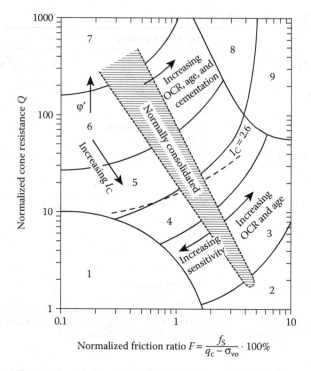

Normalized friction ratio $F = \dfrac{f_S}{q_c - \sigma_{vo}} \cdot 100\%$

Figure 3.24 CPT-based soil behavior type chart after Robertson 1990. 1 = sensitive, fine-grained, 2 = organic soils—peat, 3 = clays—silty clay to clay, 4 = silt mixtures—clayey silt to silty clay, 5 = sand mixtures—silty sand to sandy silt, 6 = sands—clean sand to silty sand, 7 = gravelly sand to dense sand, 8 = very stiff sand to clayey sand*, 9 = very stiff, fine-grained* (*heavily overconsolidated or cemented). (Redrawn from NCEER, in Youd, T.L. and Idriss, I.M. (eds.), *Proceedings of the NCEER Workshop on Evaluation of Liquefaction Resistance*, Technical Report NCEER-97-0022, 1997.)

In Figure 3.24, a soil behavior type chart is shown as developed by Robertson (1990) identifying different soil types as a function of their cone resistance and their friction ratio. The CPT friction ratio (sleeve resistance f_s divided by cone tip resistance q_c) increases with increasing fines content and plasticity of the soil. The boundaries between the different soil types of the graph can be approximated by concentric circles of the radius I_c, referred to as the soil behavior type index:

$$I_c = \sqrt{(3.47 - \log Q)^2 + (1.22 + \log F)^2} \qquad (3.11)$$

with Q and F representing the normalized cone resistance and the normalized friction ratio as follows:

$$Q = \left[\frac{(q_c - \sigma_{v0})}{p_a} \right] \cdot \left(\frac{p_a}{\sigma'_{v0}} \right)^n \qquad (3.12)$$

$$F = \left[\frac{f_s}{(q_c - \sigma_{v0})} \right] \cdot 100\% \tag{3.13}$$

I_c is calculated in an iterative process starting with $n = 1$ as exponent for clayey soils in Equations 3.11 through 3.13. If the calculated I_c value is above 2.6 the soil is generally considered too cohesive to liquefy. For safety reasons the NCEER recommends these soils to be sampled and laboratory tested for confirmation. If the computation arrives at I_c values less than 2.6 the soil is more likely granular, and calculations of Q and I_c should be repeated with $n = 0.5$ for sand. If I_c then falls below 2.6, the soil is considered nonplastic and granular and I_c can be used to determine the liquefaction potential for sand with fines contents as explained below. However, should I_c be again greater than 2.6, the soil is likely to be cohesive and plastic and I_c should be recalculated with the intermediate value of $n = 0.7$. The intermediate I_c is then used to establish the liquefaction potential of this soil, which should also be sampled and tested for verification purposes according to NCEER recommendations. Soils with a soil behavior index I_c above 2.6 would generally not liquefy but would be rather soft (region 1 in Figure 3.24) and could therefore suffer considerable deformation during an earthquake. These soils would also be considered nonliquefiable according to the so-called Chinese criteria by which liquefaction would only occur if all three of the following criteria were met (after Seed and Idriss, 1982):

1. The clay content is less than 15%.
2. The liquid limit is less than 35%.
3. The natural moisture content is less than 0.9 times the liquid limit.

Correction of the normalized cone penetration resistance $q_{c,1N}$ of these silty sands to an equivalent clean sand value (index cs) for determining the liquefaction resistance CRR from Figure 3.23 is by the following relationship:

$$(q_{c,1N})_{cs} = K_c \cdot q_{c,1N} \tag{3.14}$$

where K_c represents the grain characteristics correction factor, which can be taken from Figure 3.25 as a function of the soil behavior type index I_c as defined by Equation 3.11.

To correct the $CRR_{7.5}$ values, which are only valid for 7.5 magnitude earthquakes, as taken from Figure 3.22, as a function of normalized corrected SPT blow counts, or from Figure 3.23, for normalized corrected CPT tip resistances, this value is multiplied by the magnitude scaling factor for stresses MSF_M as proposed by the NCEER and as taken from Figure 3.26, and as given in Equation 3.15:

$$CRR_M = MSF_M \cdot CRR_{7.5} \tag{3.15}$$

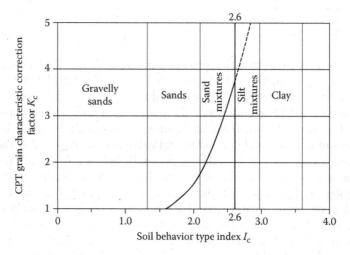

Figure 3.25 Grain-characteristic correction factor for determination of clean-sand equivalent CPT resistance. (Redrawn from NCEER, in Youd, T.L. and Idriss, I.M. (eds.), *Proceedings of the NCEER Workshop on Evaluation of Liquefaction Resistance,* Technical Report NCEER-97-0022, 1997.)

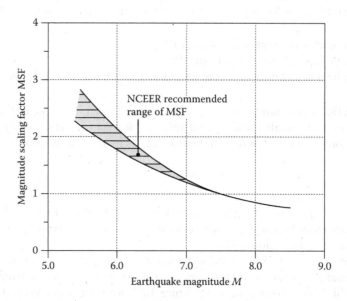

Figure 3.26 Magnitude scaling factor. (Redrawn from NCEER, in Youd, T.L. and Idriss, I.M. (eds.), *Proceedings of the NCEER Workshop on Evaluation of Liquefaction Resistance,* Technical Report NCEER-97-0022, 1997.)

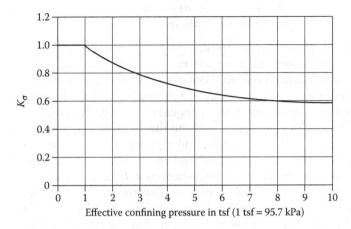

Figure 3.27 K_σ factor as a function of confining pressure. (Data from NCEER, in Youd, T.L. and Idriss, I.M. (eds.), *Proceedings of the NCEER Workshop on Evaluation of Liquefaction Resistance,* Technical Report NCEER-97-0022, 1997.)

The factor of safety (FS) against liquefaction can now be written as a function of the CSR, the CRR for 7.5 magnitude earthquakes $CRR_{7.5}$ and the magnitude scaling factor MSF as follows:

$$FS = \left(\frac{CRR_{7.5}}{CSR} \right) \cdot MSF \qquad (3.16)$$

We have seen from Figure 3.21 that the CRR values, as developed from Figures 3.22 and 3.23, are only valid for depths not exceeding 15 m. A correction factor K_σ for overburden pressures greater than 100 kPa was therefore developed. While the liquefaction resistance generally increases with increasing confining pressure, cyclic triaxial compression tests revealed that the resistance in terms of the cyclic stress ratio decreases in a nonlinear form with depths greater than 15 m. To account for this effect, the NCEER recommends the K_σ factor according to Figure 3.27 for clean and silty sands and for gravel.

For an in-depth study of the liquefaction phenomenon of saturated sand, the reader is referred to the NCEER (1997) publication and numerous other publications referenced therein.

The computational method described and proposed by the NCEER needs to be carried out for all layers of a given soil profile that are likely to liquefy during an earthquake. FS against a design earthquake can be calculated in this way, and safety profiles can be developed to establish the eventual need of soil improvement. To facilitate the calculation, particularly to avoid the cumbersome iteration procedure to find the correct soil behavior type index I_c, professional software has been developed based upon these NCEER recommendations (i.e., Shake, 2000; LiquefyPro, 2008).

3.3.3.2 Settlement estimation of sands due to earthquake shaking

Besides the potential catastrophic effects of liquefaction to structures, and their possible avoidance by soil improvement measures, the foreseeable extent of earthquake-induced deformations in granular soils is of particular interest in seismic zones. It is well-known that sand tends to compact and settle when subjected to earthquake shaking. The degree of deformation depends, of course, primarily on the initial density of the sand deposit and the earthquake magnitude, and ranges generally between less than 1% volumetric strain for dense sand and up to 5% for very loose sand.

To estimate the probable settlement as a result of earthquake shaking, Tokimatsu and Seed (1987) proposed that the calculations be divided into *saturated sand settlement* below groundwater, where generally liquefaction would occur, and *settlement for dry or unsaturated sand* above, with the sum of both representing the total earthquake-induced settlement.

It is recommended to base these calculations on SPT values and to convert any other in-situ density measurements, such as CPT results, into normalized $(N_1)_{60}$ values by using Equation 3.10 and Table 3.4. For this purpose, Figure 3.28 also provides a useful relationship (after Robertson et al., 1983) between the mean grain size D_{50} and the $q_c/(100 \cdot N)$ ratio for different types of sand.

If the soil behavior type index I_c is known, the conversion can also be done based upon the following relationship:

$$\frac{q_{c1}}{(N_1)_{60}} = 8.5 \cdot \left(\frac{1 - I_c}{4.6}\right) \tag{3.17}$$

with q_{c1} in tsf (1 tsf = 95.7 kPa).

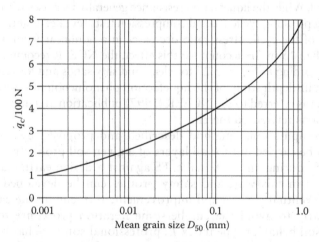

Figure 3.28 Relationship between D_{50} and $q_c/(100 \cdot N)$ ratio, q_c in kPa. (After Robertson, P.K. et al., J. Geotech. Eng., 109(11), 1449, 1983.)

The primary effect of the shaking of saturated sand is the generation of excess pore water pressure with settlement occurring as it dissipates. The time for the settlement to develop depends on the length of the drainage path and the permeability of the sand. In highly permeable sand, it may occur immediately, but in less permeable sand only after many hours.

It has been found that volume deformation decreases with increasing density of the sand and decreasing induced strain. Accordingly, the volumetric strain ε_v in saturated sand developing after liquefaction as proposed by Tokimatsu and Seed (1987) is presented in Figure 3.29 as a function of the CSR and corrected normalized SPT blow count $(N_1)_{60}$. The abscissa uses a conversion between relative density D_r and SPT N-values proposed by the same researchers, which is based upon an empirical expression which G. Meyerhof had presented in 1957 as follows:

$$D_r = 21 \cdot \sqrt{\frac{N}{\sigma_0' + 0.7}} \qquad (3.18)$$

with the effective overburden pressure σ_0' in kg/cm² (1 kg/cm² = 100 kPa).

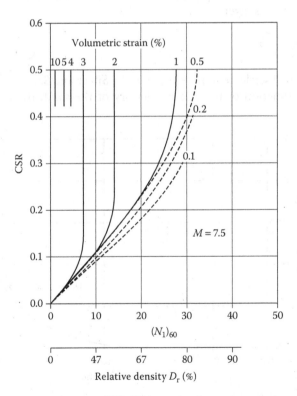

Figure 3.29 Relationship between CSR, $(N_1)_{60}$, and volumetric strain in saturated sand. (Redrawn from Tokimatsu, K. and Seed, H.B., *J. Geotech. Eng.*, ASCE, 113, 861, 1987.)

As can be seen in Figure 3.29, the earthquake-induced volumetric strain can range well over 3% in loose sand and is insignificant when dense conditions prevail. To use this figure also for earthquakes other than $M = 7.5$, the values of the cyclic stress ratio have to be multiplied by the magnitude scaling factor MFS as obtained from Figure 3.26 and using:

$$\left(\frac{\tau_{av}}{\sigma_0'}\right)_{M=M} = \left(\frac{\tau_{av}}{\sigma_0'}\right)_{M=7.5} \cdot MSF \tag{3.19}$$

The studies of Tokimatsu and Seed (1987) also revealed that, for conditions of incomplete liquefaction, only small amounts of settlements will occur, which are generally represented by volumetric strains well below 1% (dashed lines in Figure 3.29). Figure 3.30 gives the relationship between volumetric strain and normalized stress ratios below 1 (i.e., for stress levels below liquefaction stage) for saturated clean sand, resulting in a volumetric strain of just 0.1% for a stress level of 80% of that causing liquefaction.

The liquefaction induced settlement Δs of a soil layer with the thickness h is then calculated by multiplying it with the representative volumetric strain value ε_v (%) of this layer:

$$\Delta s_{sat} = \frac{\varepsilon_v}{100} \cdot h \tag{3.20}$$

For the dry soil settlement, investigations by Silver and Seed (1971) revealed that it is a function of the relative density of the sand, the magnitude of

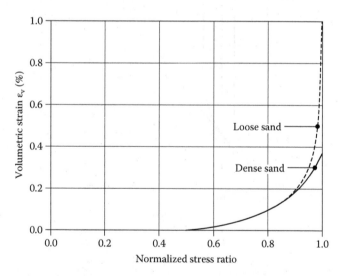

Figure 3.30 Relationship between volumetric strain and normalized stress levels for nonliquefied saturated clean sand. (Redrawn from Tokimatsu, K. and Seed, H.B., *ASCE J. Geotech. Eng.*, ASCE, 113, 861, 1987.)

the cyclic shear strain and the number of strain cycles. It is insignificantly affected by the degree of the vertical stress. At a given depth in the soil, the effective shear strain γ_{eff} can be written as

$$\gamma_{eff} = \frac{\tau_{av}}{G_{eff}} = \frac{\tau_{av}}{G_{max}(G_{eff}/G_{max})} \tag{3.21}$$

with G_{max} the shear modulus at low strain level, G_{eff} the effective shear modulus at the induced strain level, and τ_{av} the average cyclic shear stress at that depth. According to Equation 3.7, the average CSR is

$$\tau_{av} = 0.65 \cdot \frac{a_{max}}{g} \cdot \sigma_{v0} \cdot r_d \tag{3.22}$$

leading with Equation 3.21 to

$$\gamma_{eff} \cdot \frac{G_{eff}}{G_{max}} = \frac{0.65 \cdot a_{max} \cdot \sigma_{v0} \cdot r_d}{g \cdot G_{max}} \tag{3.23}$$

with

$$G_{max} = 1000 \cdot (K_2)_{max} \cdot \sqrt{\sigma'_m} \quad \text{in psf-units} \tag{3.24}$$

and

$$(K_2)_{max} = 20 \cdot (N_1)^{1/3} \tag{3.25}$$

In this way, the product $\gamma_{eff} \cdot G_{eff}/G_{max}$ can be determined for any given sand layer which is plotted as abscissa in Figure 3.31 as a function of the shear strain γ_{eff} with the confining pressure σ'_m as parameter valid for an $M = 7.5$ (15 cycles) earthquake. Tokimatsu and Seed (1987) then use the cyclic shear strain in Figure 3.32 to determine the volumetric strain ε_v for this layer.

For different earthquake magnitudes, the volumetric strain values thus developed need to be multiplied by the relevant magnitude scaling factor for strain ratios MSF2 from the fourth column in Table 3.7.

Figure 3.32 is valid for one-directional shear conditions and the ε_v values have to be multiplied by 2 to reflect multidirectional shear conditions, which normally prevail in an earthquake. The dry soil settlement of the soil layer with a thickness h is then

$$\Delta s_{dry} = \frac{2 \cdot \varepsilon_v}{100} \cdot h \tag{3.26}$$

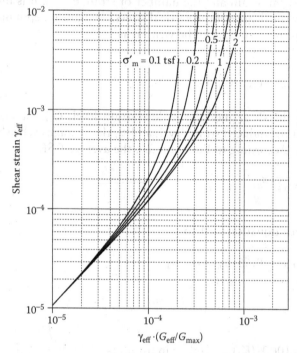

Figure 3.31 Plot to determine the induced strain in sand deposits resulting from M = 7.5 earthquake. (After Tokimatsu, K. and Seed, H.B., J. Geotech. Eng., ASCE, 113, 861, 1987.)

Table 3.7 Magnitude scaling factors for stress ratios causing liquefaction in saturated sand (third column) and for strain ratios in dry sand (fourth column)

Magnitude M	Representative stress cycles at 0.65 τ_{max}	Scaling factor MSF1 for stress ratios	Scaling factor MSF2 for strain ratios
8.5	26	0.89	1.25
7.5	15	1.0	1.0
6.75	10	1.13	0.85
6	5	1.32	0.6
5.25	2–3	1.5	0.4

Source: Tokimatsu, K. and Seed, H.B., J. Geotech. Eng., ASCE, 113, 861, 1987.

The total earthquake-induced settlement in a sand deposit is the sum of the settlements of layers below the water table and the settlements of the layers above:

$$S_{total} = \sum \Delta s_{wet} + \sum \Delta s_{dry}$$

(3.27)

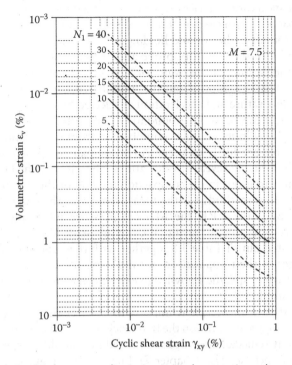

Figure 3.32 Relationship between volumetric strain, shear strain, and penetration resistance for dry sands. (Redrawn from Tokimatsu, K. and Seed, H.B., *J. Geotech. Eng.*, ASCE, 113, 861, 1987.)

3.4 QUALITY CONTROL AND TESTING

Once the design of the vibro compaction project has determined the geometrical extent of the treatment, together with the probe spacing necessary to obtain the required density using the method as described in Section 3.3.1, a suitable quality assurance program needs to be developed. The aim is to ensure that the operating parameters so obtained are observed when executing the works by repeating it for each vibro compaction probe. Such a program will have to include all elements that are essential for a well-organized special foundation site. For the vibro compaction process, water management is a very important issue. It has to be channeled away from the work position into settling ponds before it is released, normally without further treatment, into the available recipient. Similarly important is the well-organized transportation of the backfill material to the working area and its stockpiling on-site, when necessary. Continuous measurement of the added material, together with site surface level measurements before and after compaction, will help to evaluate compaction success after its execution.

The latest monitoring devices record—as a function of real time for each vibro compaction probe—penetration depth, energy consumption of the vibrator motor, lifting height and holding time for each step, total execution time, verticality of the vibratory probe and, if found necessary, pressure and quantity of the flushing media, water or air, used. These devices can also provide additional information about the working condition of the vibrator in displaying temperature and acceleration at selected locations inside the vibrator in use (Figure 3.33). As the ground to be compacted is in general a natural deposit, variations in its granular composition from the location of the trial compaction area are likely to occur. An accompanying postcompaction test program will normally reveal any deviations from the specified requirements, deriving from variations in the compaction method or the soil properties.

Suitable testing methods for the measurement of the compaction result, such as the SPT, the CPT, the Menard pressure meter test (MPT), the shear wave velocity measurement (V_s), the full-scale load test (LT), and the DDM, are listed out in Table 3.8 together with their specific advantages or disadvantages of their application in sand deposits. Under special circumstances, and when specific knowledge exists, other in-situ testing methods may also be used.

Before starting a project, it is important to select the appropriate testing method and to agree not only on the frequency of pre and postcompaction testing but also any remedial measures and procedures should the specified density not have been met (see also Chapter 7). Engineering judgment is required to determine the need and extent of recompaction that generally follows only after the deficiency was verified by a second test in the vicinity of the failed test.

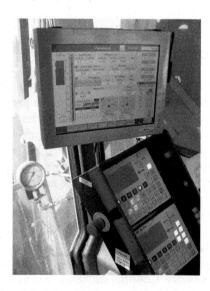

Figure 3.33 Display of multifunctional vibro compaction control panel. (Courtesy of Keller Group plc, London, UK.)

Table 3.8 Suitable testing methods to measure compaction results in sand for the assessment of the liquefaction potential

Type of test	Available data	Repeatability	Depth range	Measures index or property	Can test provide soil sample?	Detection of soil variability	Cost
SPT	Abundant	Poor to good	Deep	Index	Yes	Good	Low
BPT	Sparse	Poor	Deep	Index	No	Fair	Low
CPT	Abundant	Very poor	Deep	Index	No	Very good	Low
MPT	Sparse	Poor	Medium	Property	No	Fair	Moderate
V_s	Limited	Good	Deep	Property	No	Fair	Low
LT	Limited	Very good	Low	Property	No	Poor	High
DDM	Limited	Poor	Low	Property	Yes	Very good	Very high

Source: NCEER, in Youd, T.L. and Idriss, I.M. (eds), *Proceedings of the NCEER Workshop on Evaluation of Liquefaction Resistance*, Technical Report NCEER-97-0022, 1997.

Notes: SPT, standard penetration test; BPT, Becker penetration test; CPT, cone penetration test; MPT, Menard pressure meter test; V_s, shear wave velocity measurement; LT, load test; DDM, direct density measurement.

It is also very important that all pore water pressure increases created by the vibratory motion in the treated soil are allowed to fully dissipate before any postcompaction testing commences. Full relief of the excess pore water pressure in the soil has normally occurred about one week after compaction. In cases of doubt, it is better to wait for a longer period or to measure the pore water pressure development over time directly. To avoid any adverse influence on the postcompaction testing, this criterion should be adhered to for a zone of about 30 m around the testing area where no simultaneous compaction work should be carried out.

In special, not yet fully understood, circumstances, a so-called aging effect can occur in sand whereby the strength properties of the compacted sand continue to increase significantly for up to several weeks after compaction (Mitchell et al., 1984; Massarsch, 1991; Schmertmann, 1991). It is therefore recommended to perform field trials at the beginning of larger projects with in-situ testing to start at different time intervals after compaction. By this way, the significance of any potential time effect on the posttreatment test results can be assessed, and the appropriate conclusions may be drawn. Increases in cone resistance q_c of up to 100% have been reported 10 weeks after the first postcompaction test performed one week after completion of deep compaction.

According to an analysis made by Charlie et al. (1992, from Lunne et al., 1997), covering projects that represent a wide range of geological and climatic conditions, the variation of the cone resistance q_c with time can be expressed by an empirical relationship as follows:

$$(q_c)_N = (1 + K \log N)(q_c)_1 \tag{3.28}$$

with

N is the number of weeks

K is an empirical constant, dependent on the average air temperature T measured in °C according to Equation 3.29

$$K = 0.04 \cdot 10^{0.5 \cdot T}$$ (3.29)

$(q_c)_1$ is the normalized q_c after 1 week

$(q_c)_N$ is the normalized q_c after N weeks

The temperature-dependent relationship indicates that chemical reactions may support intergranular chemical bonds developing with time between particles and being responsible for this strength increase.

3.5 SUITABLE SOILS AND METHOD LIMITATIONS

It can be regarded a prerequisite that the need for compaction of loose soils or soils with varying density is established by a site investigation. As a rule, soils suitable for vibro compaction should be granular in nature with a negligibly low cohesion or plasticity. For a quick assessment of the suitability of granular soils for treatment by vibro compaction, Degen (1997b) has proposed a first approach based upon the USCS together with useful comments on the compaction method as shown in Table 3.9. Accordingly, suitable soils are generally sand and gravel. Ideally their silt content (grain size below 0.06 mm) should be below 10%. Larger percentages of silt, and any clay content, will considerably obstruct the compaction process, if not prevent it totally.

Figure 3.13 shows application limits in a grain size distribution graph where area B represents soils that are ideally suited for vibro compaction

Table 3.9 Suitability assessment of granular soils for vibro compaction

Soil type	USCS	Comment on suitability for vibro compaction
Gravel, well graded	GW	Well suited for vibro compaction, potential penetration difficulties with less powerful machines
Gravel, poorly graded	GP	If $D_{60}/D_{10} \leq 2$ compaction only marginal (trial compaction recommended)
Gravel, silty or clayey	GM, GC	Compaction not possible if clay content >2% and silt content >10%
Sand, well graded	SW	Ideally suited
Sand, poorly graded	SP	If $D_{60}/D_{10} \leq 2$ compaction only marginal (trial compaction recommended)
Sand, silty	SM	Compaction inhibited if silt content >8%
Sand, clayey	SC	Compaction inhibited if clay content >2%

Source: Modified from Degen, W., Vibroflotation Ground Improvement (unpublished), 1997b.

with a content of fines below 10%. The soils represented by area A are very well-compactable, but an increasing amount of coarse gravel and cobbles also increases permeability leading to a total loss of water at about $k = 10^{-2}$ m/s. This may obstruct penetration of the depth vibrator to such an extent that the desired depth cannot be reached in some or all parts of the site. In cases like this, where the process may become uneconomic or even impossible, to execute a precontract trial is recommended.

Such trials may also be advisable for soils falling into area C, where vibro compaction is still possible but only with considerably extended compaction time. Speed and effectiveness of the vibro compaction process depend largely on the permeability, k, of the sand. At a permeability below 10^{-3} m/s, penetration of the vibrator will increasingly be slowed down. While for soils lying in areas A and B the necessary backfill material for compensation of the surface settlement resulting from the compaction process itself can be taken from the surface, imported coarser (i.e., more suitable) backfill material is necessary when the soils to be compacted fall into area C. Soils falling completely or partially in area D cannot be compacted by the deep vibratory process. Vibro replacement stone columns (see Chapter 4) or other foundation measures may become necessary in such a case.

The boundaries described above have all been established empirically over many years of application, and it must be kept in mind that specialist contractors often rely on considerable knowledge and experience when it comes to borderline applications. In this context, the choice of the appropriate vibrator with specific characteristics can be decisive for the execution of vibro compaction. A vibrator with particularly good penetration characteristics may be advisable for soils in area A, while vibrators compacting at lower frequencies (below 30 Hz) will perform better in conditions of area C—hence the development of depth vibrators with variable frequency that combine both characteristics within one machine.

In addition to the particle size distribution of the soil, static CPTs may also be used to establish compaction suitability of soils. Figure 3.34 shows an empirical relationship proposed by Massarsch (1994) between cone penetration resistance and friction ratio, and defines the zone of compactable soils to fall within friction ratios below 1% at a point resistance of at least 3 MPa. It also indicates a zone for soils that are only marginally compactable with friction ratios between 1% and 1.5% at q_c values between 1 and 3 MPa.

A suitability number (SN) based on grain size distribution has been proposed by Brown (1977), according to Equation 3.30:

$$SN = 1.7 \left(\frac{3}{D_{50}^2} + \frac{1}{D_{20}^2} + \frac{1}{D_{10}^2} \right)^{0.5} \tag{3.30}$$

D_{50}, D_{20}, and D_{10} are grain size diameters in millimeters at 50%, 20%, and 10% passing in a grain size distribution curve. It is suggested that a low SN is better suited for vibro compaction than a high number of 40–50, above

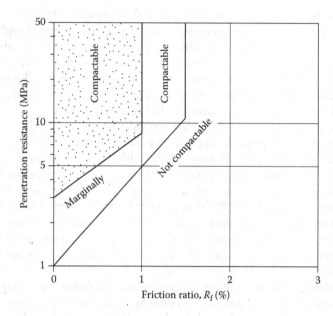

Figure 3.34 Compactable soils as a function of cone penetration resistance and friction ratio. (After Massarsch, K.R., Design aspects of deep vibratory compaction, in *Proceedings of Seminar on Ground Improvement Methods*, Geotechnical Division, Hong Kong Institution of Engineers, Hong Kong, China, May 19, 1994, pp. 61–74.)

which the ground is unsuitable, and that quicker compaction is achieved with lower numbers.

Very steep grain size distribution curves, with D_{60}/D_{10} below 2, are a strong indicator that vibro compaction may only be marginally possible if not inhibited completely. A compaction test in the laboratory, or a field trial, will determine the suitability of such soils when no local experience exists.

The above approaches provide reasonable guidelines, but permeability plays an important role in speed and effectiveness of vibro compaction. With decreasing permeability below $k = 10^{-5}$ m/s, compaction is increasingly inhibited, while very high permeability in excess of 10^{-2} m/s may slow penetration of the vibrator as a result of the loss of water. In this instance, where jetting water is normally of limited use, more powerful machines operating at higher frequencies, in excess of 50 Hz, are more suited and are capable of penetrating even highly permeable gravel and cobble. In this category are applications where brick demolition rubble is regularly compacted dry in regenerating derelict industrial city centers in the North of England to at least the depth of the former basements.

Where shallow treatment depths below 1.5 m are concerned, standard surface compaction methods such as vibratory roller compaction or heavy tamping tend to be more economical. This is especially so since the upper 1.0 m of sand is generally not well-compacted enough by vibro compaction requiring either removal or subsequent surface densification by standard methods.

Very deep compactions, well in excess of 50 m, are known to have been executed successfully in exceptional cases, either to mitigate earthquake risks, to improve stability hazards of slopes, or in other similar cases. In these instances, besides very heavy cranes, considerable experience is required in safely operating the extended depth vibrators. Although deep vibratory compaction is a very versatile method, in-situ trial compactions may be advisable when unknown territory is entered.

The need to compact newly dredged sand in reclamation areas for the safe foundation of quay walls, berths and similar structures often require densification of the foundation soils through deep water. This can be achieved from floating barges and pontoons or, in shallower water, from special platforms. In offshore applications, the critical problem will always be the accurate setting of the compaction locations, which can today be mastered by satellite-supported positioning and inclinometer measurements. As a result of the complex surveying problems, the use of multiple depth vibrators (more than four units have been used) may be advisable. It is only heavy seas or swells and strong water currents that may prevent a safe, secure performance of the vibro compaction process until a friendlier environment prevails again.

Deficiencies in the material grading can also affect penetration of the vibrator and successful compaction of sandy soils. A layer of cobbles or gap-graded material of fine uniform sand with floating cobbles may prove difficult for the compaction process; the cobbles can accumulate in front of the vibrator point and obstruct further penetration. The use of air as the flushing medium or changing the vibrator frequency can help to alleviate the situation.

However, particle size distribution, in-situ density measurement, and permeability are not always sufficient to define the suitability for vibratory deep compaction, nor are the in-situ measurements always easy to interpret. The mineral composition and specific gravity of the sand deposit can greatly influence the result of indirect density measurements. In this context, and in contrast to silica sand, in carbonate sands containing considerable amounts of shell debris the evaluation of CPT results cannot be translated easily into relative densities.

Various researchers have found that even relatively small percentages of shell debris (10%–20%) (Vesic, 1965; Cudmani, 2001) in silica sand have considerable influences on the CPT point resistance at the same density. Bellotti and Jamiolkowski (1991) found a linear relationship between the q_c(silica)/q_c(shell) ratio and the relative density D_r as given in Equation 3.31. Ratios between 1.5 and 3 have been reported in special publications.

$$\frac{q_c\,(\text{silica})}{q_c\,(\text{shells})} = 1 + 0.015(D_r - 20) \qquad (3.31)$$

It is therefore recommended to correlate the CPT results with direct in-situ density measurements or to conduct special laboratory calibration tests between density and penetration resistance, whenever the carbonate content resulting from shell debris is in excess of 10%, unless enough experience and data exist for the referenced soil.

Wehr (2005a) presented a shell correlation factor f_s as a function of the relative density D_r for Dubai sand (carbonate content in excess of 90%, $D_{60}/D_{10} = 3$) by which the CPT results measured in the calcareous sand need to be multiplied to arrive at corresponding values for silicate sand:

$$f_s = \frac{q_c \text{(silica)}}{q_c \text{(shell)}} = 0.0046 \, D_r + 1.3629 \tag{3.32}$$

The study revealed that the influence of the in-situ density was in all tests so dominant compared with the overburden pressure that the latter could be neglected. Accordingly this correlation factor ranges between 1.4 and 1.8 and amounts to 1.64 for a relative density of 60%. It was stated that the values found represented lower limits since all shell particles larger than 4 mm in diameter were removed from the test specimens representing about 8% of the total weight.

In general it is always recommended, in addition to keeping an accurate record of all imported backfill material, to also consider the measured surface settlement after vibro compaction as an indicator and as supporting evidence of the increase in density achieved comparing the pre and post-compaction volumes.

When interbedded sand layers with cohesive soils above and below required compaction, it is often found that the result of the standard deep compaction is below expected levels. The reduced compaction is especially noticeable at the upper and lower interface in the granular soil with the cohesive material. It is surmised that the vibratory motions are dampened to a certain extent by these cohesive layers. The proper functioning of the sand backfill is important in these cases to compensate the volume reduction in the sand layer which will come to the fore at the upper interface. An extended compaction time and, if necessary, also reduced compaction probe spacing will help to avoid this problem.

Natural cementation of sand can be found in many parts of the world. It occurs predominantly in arid climatic conditions in calcareous sand deposited above the water table. Cementation can also occur when hydrous silicates or iron oxides act as a kind of glue at the contact point between the sand particles. Cementation may indicate a greater density or strength than actually exists and in-situ density measurements and their standard correlation may result in misleading interpretations. Weakly cemented soils with actual low relative densities may respond to vibratory compaction by collapsing and yet producing lower post- than pretreatment test results, although the compacted material is now in a much more stable condition. Strong cementation may also be an obstacle that cannot be overcome by the vibratory treatment without the help of preboring. To conclude, it is important to know if particle cementation exists at a site where vibro compaction is envisaged. Only then can appropriate measures be designed for its in-situ improvement.

3.6 CASE HISTORIES

3.6.1 Vibro compaction for a land reclamation project

In early 2000, the island state of Singapore embarked on another major reclamation project, for which it is renowned and by which it seeks to overcome the chronic land scarcity, a major obstacle for the development of the country. Led by the Jurong Town Corporation, a much needed industrial space was developed at Tuas, which is located toward the western tip of Singapore Island.

The vibro compaction method was chosen to safeguard the stability of the hidden dyke that defines the boundaries of the reclamation area (Figure 3.35). It is composed of a well-graded, gravelly, fine to coarse sand with an in-situ relative density after dredging of only 35% (Figure 3.36). Specifications required a minimum cone tip resistance measured by the CPT method of

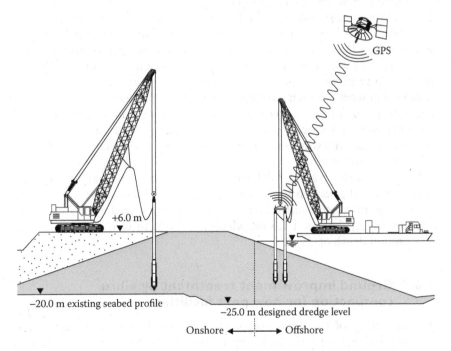

Figure 3.35 Vibro compaction at hidden dyke. (Courtesy of Keller Group plc, London, UK.)

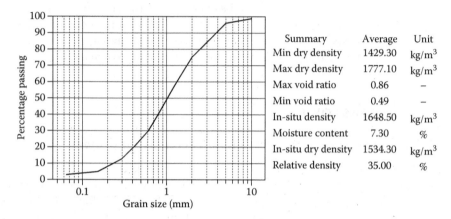

Figure 3.36 Typical grain size distribution and properties of sand.

4 MPa between 0 and 2 m depth, of 6 MPa between 2 and 8 m, and of 8 MPa below. Vibro compaction had to be carried out up to a maximum depth of 35 m.

The chosen compaction grid pattern, an equilateral triangular grid of 4 m, was established together with other operational details, such as the amount of water used and compaction time, in a field trial. In total, 22.4 million m³ of sand were finally compacted, 67% on the dry from the land side, 33% offshore as a marine operation. Two heavy crawler cranes were used on land each carrying a single S300 Keller depth vibrator with characteristics according to Table 3.1. For the marine work, a 200-ton crane with twin S300 vibrators was mounted on a barge to carry out the compaction work. In such a configuration, work was conducted in day and night shifts for six days a week, compacting approximately 1 million m³ of sand per month. To locate the compaction points with sufficient accuracy over water, the GPS was used with antennas mounted directly on top of the vibrator extension tubes.

Quality control was by CPT carried out seven days after the completion of a specific section of the works. Figure 3.37 shows typical pre and postcompaction test results. Surface settlement measured after vibro compaction averaged around 1.5 m, which is equivalent to 6% of the average compaction depth of 25 m.

3.6.2 Ground improvement treatment by vibro compaction for new port facilities

The versatility of the vibro compaction process in solving special foundation problems is particularly appreciated where harbor construction is concerned. The need to increase the density of dredged sand fill behind or

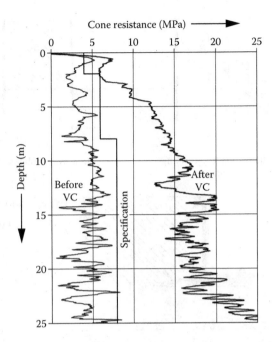

Cone resistance (MPa) ⟶

Figure 3.37 Pre- and postcompaction test results. VC, vibro compaction. (Courtesy of Keller Group plc, London, UK.)

below quay walls or inside and outside coffer dams is a typical requirement that can best be achieved by this method (Figure 3.38).

The quayside unloading and breakwater facilities of the Merak harbor in West Java, Indonesia, serving the delivery of material for a pulp and paper manufacturing plant, are shown in Figure 3.39 with a typical cross section through the quay wall. The fill material is a fine to medium sand with basically no fines with a $D_{60}/D_{10} = 4$. About 50% of its grains are silica sand, 20% shells and 30% other fragments with an overall specific gravity of 2.82.

Structural stability and the reduction of the liquefaction potential required the hydraulic sand fill in the areas shown to be compacted to a relative density of 70%. In total, some 1.3 million m³ of sand fill needed treatment to up to 25 m depth. The layout of the compaction probes was chosen after a trial compaction before the start of the works as a 3 m square grid.

Quality control consisted of continuous measurement and recording of vibrator motor current over depth as a function of real time, together with postcompaction tests carried out as CPTs, 10 days after completion of a defined section of the works not exceeding 750 m². The specified density criterion of 70% relative density was generally exceeded in this well-compactable sand material. In addition, surface settlements induced by the

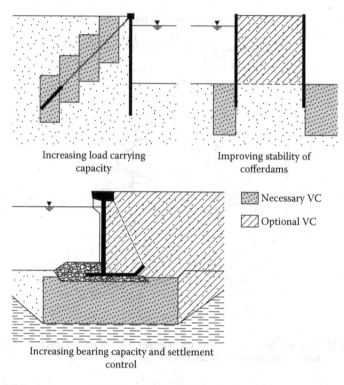

Increasing load carrying
capacity

Improving stability of
cofferdams

[:::] Necessary VC

[///] Optional VC

Increasing bearing capacity and settlement
control

Figure 3.38 Typical examples for ground improvement by vibro compaction in harbor construction. VC, vibro compaction.

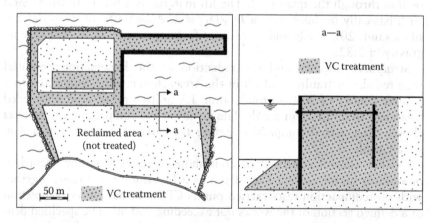

Figure 3.39 Plan of quay walls and breakwater with treatment area and cross section through quay wall. VC, vibro compaction. (Courtesy of Keller Group plc, London, UK.)

compaction were measured for each works section. They reached on average 7% of the compaction depth with maximum values in areas of particularly clean sand of more than 10%.

3.6.3 Vibro compaction field trial in calcareous sand

On a major prestigious development project in the Middle East that was realized between 2006 and 2009, different specialist contractors were engaged to allow the developer to follow a very tight time schedule for his project, in particular for the dredging operation and the following compaction works. Consequently, a variety of depth vibrators were employed which were all tested on-site prior to the start of any contractual work. The tests were designed to establish the required vibro probe grid spacing to achieve the specified density as was laid down in a CPT target curve according to Figure 3.42. Accordingly ground improvement had to be designed and conducted to ensure an allowable bearing pressure of 150 kN/m² at 1 m below ground surface, total settlement (from static loads and liquefaction induced) not to exceed 25 mm at a maximum distortion of 1:500, and to eliminate the liquefaction risk of a design earthquake not exceeding a magnitude of 6 with a maximum ground acceleration of 0.15 g at bed rock level.

The dredged sand material was a light brownish grey to grey, shelly to very shelly, fine to medium carbonate sand (its $CaCO_3$ content was more than 90%) with many broken shell fragments smaller than 20 mm. The sand was generally only loose to medium dense. The average depth of the man-made deposit was about 16 m, which is followed by the natural ground, consisting of cap rock and weathered limestone. Ground surface was generally 3 m above mean water level of the adjacent sea. Figure 3.40 gives the band of grain size distribution curves which is representative of the project area. Its fines content (grain size less than 0.063 mm) was generally

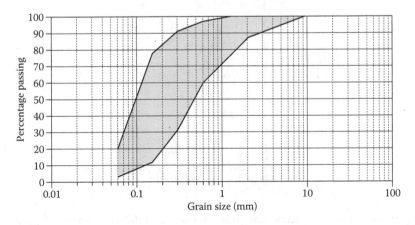

Figure 3.40 Representative grain size distributions.

less than 10% and its clay content below 2%. The mean grain size D_{50} was 0.1–0.5 mm with a maximum grain size of 10 mm.

The following describes the details of field trials using twin Keller S700 vibrators with characteristics according to Table 3.1. The total weight of the 32.1-m long set-up, consisting of a tandem beam, 500 kN pulley head, extension tubes, water and power inlets and vibrators, was 31.1 tons. The safe operation of this special equipment required the use of a 90-ton crawler crane with a boom length of ca. 37 m (see Figure 3.41).

Vibro compaction probes were installed in four trial areas of 30 × 30 m each, using triangular grid spacings of 4.2, 4.5, 4.7, and 5.0 m, respectively. The following working sequence was adopted during the trials, ultimately also forming the method for the construction work.

Probe penetration started with placing the vibrators over the probe location and opening the bottom water jets. In this phase, the depth vibrators were set to rotate at 2000 rpm. During penetration into the ground, the upper side jets were opened at a depth of about 4 m. The bottom jets were generally closed again just before reaching final depth. Now the vibrators were switched to compaction mode with an operating frequency of 1300 rpm and were held in this deepest position for 1 min or until a current consumption of 420 A was reached, whichever occurred first. The vibrators were then withdrawn in lifts of 0.75 m. After each lift, the vibrators were held for 40 s or until 420 A was reached. The craters that developed around the vibrators as compaction proceeded were filled with sand collected at ground level whenever the crater depth reached about 2 m. When approaching ground

Figure 3.41 Vibro compaction equipment set up for S700 twin vibrator operation. (Courtesy of Keller Group plc, London, UK.)

Table 3.10 Trial compaction details to obtain necessary grid spacing

Area	A	B	C	D
Number of probes	60	50	46	42
Grid spacing (triangular) (m)	4.2	4.5	4.7	5.0
Depth (average) (m)	13.96	13.82	13.90	13.91
Penetration time (min)	<4	<4	<6	<4
Compaction steps of 0.75 m (nos.)	18	18	18	18
Total compaction time (min)	16	17	17	19
Completion time per probe (min)	20	21	23	23
Subsidence (m)	1.170	1.050	0.925	0.906
Relative subsidence (%)	8.4	7.6	6.6	6.5

surface, the vibrators were switched back to penetration mode at 4 m to avoid twisting of the hanger slings. Table 3.10 gives further details of the trials performed.

Postcompaction ground levels were also taken to calculate the soil subsidence resulting from deep compaction as an additional indicator of soil densification. After a minimum of two weeks, CPT started to measure the densities achieved, for comparison with the postcompaction criteria of the contract. As explained in Chapter 7, a special procedure was adopted—in this case reflecting the calcareous nature of the sand—consisting of a pair of postcompaction CPTs at the one-third and at the midpoint of the triangular grid, as shown in Figure 7.1 with the weighted, rolling average calculated according to Equation 7.1. From the comparison of the postcompaction CPT curves thus established for the different grid patterns with the CPT target curve, the 4.5 m grid was finally chosen for the contract (see Figure 3.42).

As can be seen from Figure 3.42, the specified cone resistance was not always achieved in the upper 2–3 m. It was therefore decided to perform a surface compaction employing the novel impact rolling compaction method at least 2 weeks before contract verification testing was allowed to commence. Figure 3.43 shows an impact roller in operation. According to Avalle (2007), impact rolling is an efficient and highly productive surface compaction method requiring a careful design before, and a high degree of quality control during, its execution.

From the total compacted volume of 141 million m³, more than 30 million m³ of sand fill was compacted using Keller S700 depth vibrators by adopting the method described. On average, a shift production of 14,500 m³ was achieved for twin vibrator operation. The specified compaction criterion was achieved without the need for retesting and recompaction for 68% of the total area. For 17% of the area, the performance line was met after retesting. For 9%, recompaction was necessary to achieve the specified density, while for only 6% of the total area, verification of the density had to be done by a liquefaction analysis based on the local conditions encountered.

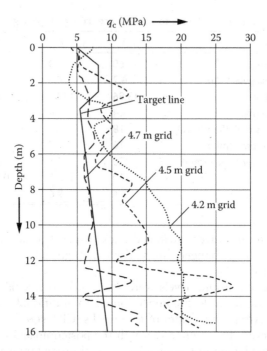

Figure 3.42 Comparison of average post-CPT results for different grid patterns with target curve.

Figure 3.43 Impact roller in operation. (Courtesy of Keller Group plc, London, UK.)

3.6.4 Foundation of a fuel oil tank farm

For the extension of a large power and desalination plant in the Middle East, 16 fuel oil storage tanks were to be constructed in 2008. The soil conditions at the site typically consist of an approximately 10 m thick layer of sand which rests on relatively incompressible bedrock made up of silt and limestone. The sand deposit consists of calcareous material of loose to medium density, generally clean to slightly silty, in places also gravelly, but its density was not sufficient to carry the high load of 250 kN/m² of the steel tanks of equal diameter of 39 m which were of the fixed roof type.

Foundation design required densification of the sand layer below the tanks together with a 5 m wide strip outside to 80% relative density, as represented by specified cone resistance target curve with q_c-values of up to 10 MPa at 3 m depth, of 15 MPa between 3 and 5 m, and 20 MPa between 5 and 7 m. Maximum settlements were restricted to 50 mm at the center of the tanks and 25 mm below the ring beam supporting the tank shell.

Trial compactions were carried out at a representative location within the future tank farm using Keller S300 vibrators on three different triangular grids with probe spacings of 2.20, 2.50, and 2.75 m. Generally all grids showed a significant increase of the CPT cone resistance after vibro compaction, and a considerable ground surface subsidence of about 8%. Although all grids met the target curve within the upper 7 m, CPT tip resistance fell below in a layer situated just a few meters above bedrock of about 1 m thickness where the friction ratio rose generally above 0.5% at a cone resistance below 5 MPa, representing the presence of silty sand, which generally can only be marginally compacted (see Figure 3.34). Therefore, conservatively the 2.20 m grid was chosen for the vibro compaction work for all tanks. In addition, a row of 0.90 m diameter vibro replacement stone columns were installed below the tank shell at center-to-center distances of 2 m. These columns extended 6.50 m into compacted sand.

Settlement performance of the chosen design was demonstrated by a finite element analysis based upon the constraint modulus of the sand as developed from the postcompaction CPT results. For the maximum settlement in the tank center 44 mm were computed, and below the tank shell 24 mm. About 10% of these settlements derived from the bedrock material below the zone of soil improvement.

Contract vibro compaction work was carried out using a pair of S300 vibrators suspended from a standard crawler crane. The stone columns construction followed, employing the wet vibro replacement method using the same depth vibrator. Surface subsidence was measured after vibro compaction for each tank ranging between 6.7% and 13.3% depending on the presence of silty sand layers, with an average of 8.5%. Contract conditions required the relatively large number of 21 postcompaction CPTs to be carried out for each tank for quality control purposes, which is equivalent to one test for every 100 m² of compacted ground. Figure 3.44 shows average CPT curves developed for each tank, which indicate that the 80% relative density target was generally met, except in cases where the presence of a high silt content in the sand in zones close to bedrock reduced or hampered the effect of vibro compaction and where vibro stone columns were additionally installed.

In addition to the CPTs, three zone load tests were also carried out using a prefabricated reinforced concrete footing measuring 2 × 2 × 0.6 m placed in the center of four vibro compaction probes. The load tests were set up and performed in accordance with ASTM D1194, employing a loading platform

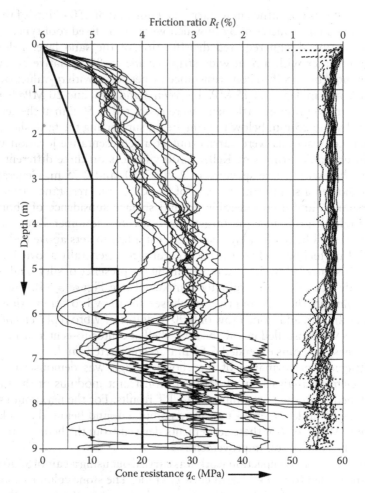

Figure 3.44 Posttreatment cone penetration tests (each curve representing the average of 21 CPTs) and CPT target line.

with a total dead load of 2200 kN consisting of concrete blocks. The load was applied in equal increments of 25% of the design footing pressure of 250 kPa up to a maximum load equal to 500 kPa. Figure 3.45 gives the pressure versus settlement curve for one of the load tests, and Table 3.11 shows the recorded settlements at design and twice design bearing pressure for all three load tests.

The relatively high stiffness as derived from the load tests indicates a better compaction than expected. When using a constrained modulus of 145 MN/m², which was measured in load test 1, for a settlement calculation, the tank centers will settle less than 20 mm, which compares favorably with the finite element analysis mentioned above in which the stiffness employed was derived from a correlation with post-CPT tip resistances.

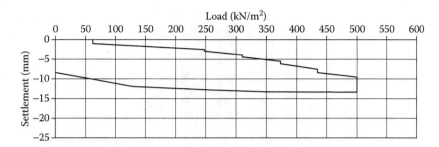

Figure 3.45 Typical pressure–settlement curve for a 2 × 2 m footing on vibro compacted sand. (Courtesy of Keller Group plc, London, UK.)

Table 3.11 Settlements and moduli as obtained from load tests

| | $p = 250 \text{ kN/m}^2$ | | $Kp = 500 \text{ kN/m}^2$ |
Load test	Settlement s (mm)	Modulus* E (MN/m²)[n]	Settlement s (mm)
1	3.06	145	13.27
2	2.54	173	9.44
3	2.55	173	6.81

* $E = 0.88 \cdot p \cdot B/s$ with footing width $B = 2$ m.

3.6.5 Liquefaction evaluation of CPT data after vibro compaction and stone column treatment

The ground improvement program for a Southern California church building was completed in late 2007 and addressed the liquefaction-induced settlement under the site design earthquake. To accomplish the site liquefaction mitigation, the soils were densified, drained, and reinforced. In general, procedures for the liquefaction mitigation with vibro stone columns were in accordance with the methods presented by Baez and Martin (1993) and Baez (1995). The stone column program consisted of a primary spacing of 19 ft (5.8 m) center-to-center and secondary columns at the midpoint of the square grid to achieve a replacement ratio of 14.1%. Figure 3.46 shows a typical soil profile and the stone columns. Totally 416 primary stone columns and 380 secondary columns were installed at the site with an average column diameter of 3 ft (0.9 m) and a maximum column length of 48 ft (14.6 m).

A liquefaction evaluation was performed on the posttreatment CPT data for this project. The following outlines the analytical approach and compares CPTs before and after the stone column treatment.

Liquefaction-induced settlement analyses were based on Tokimatsu and Seed (1987) procedures and NCEER (1997) for site liquefaction evaluation. To be conservative, the soil thin layer connection was not used. The site is very close to an active fault and has a design earthquake magnitude of 6.6 and a peak ground surface acceleration of 0.69 g. The site design water table depth

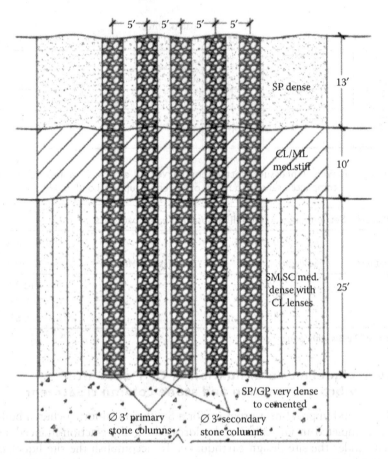

Figure 3.46 Typical soil profile with stone columns under the church building, all dimensions in feet. (Courtesy of Hayward Baker, Hanover, MD.)

is 12.5 ft (3.8 m) below ground surface. A factor of safety of 1.1 was used in the liquefaction analysis and its settlement calculation. The stated procedures were developed as a function of the penetration resistance. Results of the calculations at location of pre-CPT and post-CPT are plotted in Figure 3.47. As shown, the stone column treatment significantly increased the cone tip resistance. However, the increase of the tip resistance caused the soil behavior type index I_c to decrease when compared with pretreatment values.

Since the posttreatment I_c value decreases, the fines content calculated from post-CPT seems lower than the pre-CPT, therefore causing higher posttreatment liquefaction-induced settlement based on the Tokimatsu–Seed procedure.

Obviously, the vibro stone column treatment cannot reduce the soil's fines content. To correct the fines content calculation error, the calculation was modified, maintaining a similar fines content as the pre-CPT data (shown in Figure 3.48) and the liquefaction-induced settlement was calculated

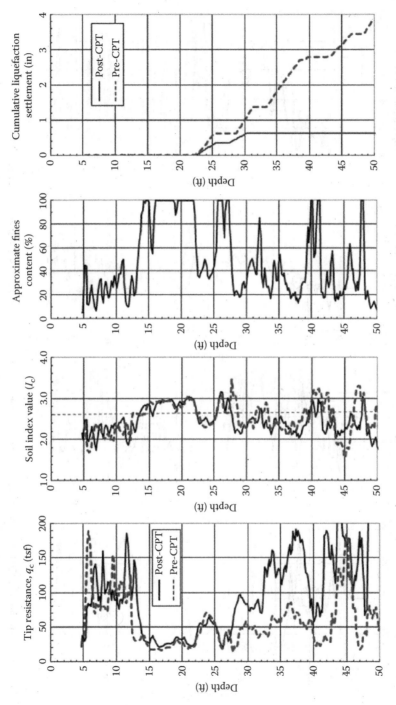

Figure 3.47 Comparison between pre and posttreatment CPT before I_c correction. (Courtesy of Hayward Baker, Hanover, MD.)

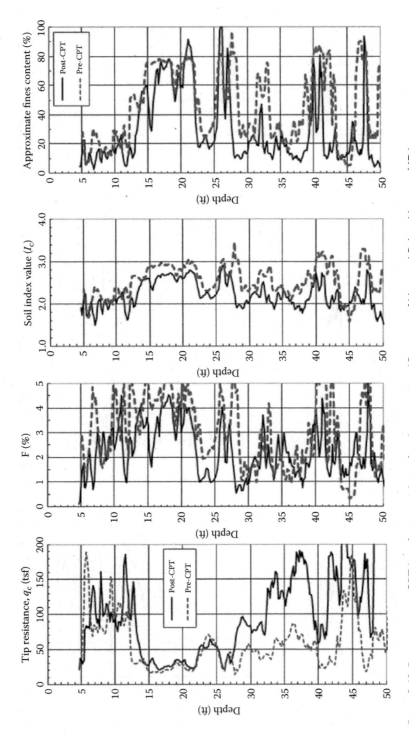

Figure 3.48 Posttreatment CPT liquefaction analysis after I_c correction. (Courtesy of Hayward Baker, Hanover, MD.)

accordingly. This I_c correction reduced about 1 in. (25 mm) calculated liquefaction-induced settlement, compared with these before I_c correction. The posttreatment CPTs were performed two weeks after stone column installation at the ground surface elevation 4.5 ft (1.4 m) below original grade. In Figures 3.47 and 3.48, the CPT depths are therefore adjusted to match the pre- and post-CPT elevations.

Based on the local building code, the site liquefaction analysis should be based on a 200 years flood water table depth, which is 12.5 ft (3.8 m) below the ground improvement working elevation. The real groundwater table depth during production was about 32.5 ft (9.9 m). To assist the bottom feed, S300 vibro probe penetration through the unsaturated medium stiff to very stiff clay layers present on-site, a Bauer BG-24 drill rig was used to pre-drill the top 30 ft (9.1 m) with a 2 ft (0.6 m) diameter auger.

The effectiveness of vibro compaction is directly related to the soil type. Figure 3.49 shows the comparison of normalized CPT tip resistance between pre and posttreatment as a function of I_c and the calculated soil's fines content, with the I_c values being the corrected values according to Figure 3.48.

It must be emphasized here that the soil liquefaction and vibro compaction share the same mechanism, for example, the sand densification under cyclic shear stress. The soils near the vibrator experienced an extremely strong *artificial earthquake*. Therefore, the soil liquefaction screening criteria can be used directly to evaluate the vibro densification effectiveness. In soils with a soil behavior type index I_c in excess of 2.6 and considered as nonliquefiable, the vibro stone column treatment only has minimal effect in terms of post-CPT tip resistance, as shown in Figure 3.49a.

Traditionally, geotechnical engineers use the soil's fines content as an indicator of its suitability for vibro compaction, especially in silty sands

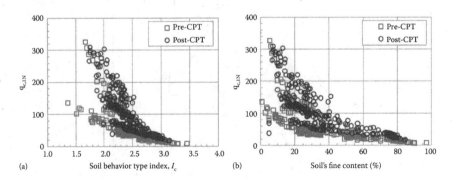

Figure 3.49 The effectiveness of vibro densification (a) related to I_c and (b) the soil fines content. (Courtesy of Hayward Baker, Hanover, MD.)

with SPT sampling and laboratory gradation test results. However, the clay content in sands has more impact on their densification and lique-faction behavior than the mere percentage of fines passing the 200# sieve (0.07 mm). Although less expressive, Figure 3.49b indicates that sands with a soil fines content interpolated from CPT in excess of 30% do not respond anymore to vibratory compaction.

3.6.6 Trial compaction in quartz sand to establish compaction probe spacing

The foundation of a multistory office building in Berlin required densifica-tion of the underlying subsoil existing of loose medium silicate sand (SW) of glacial origin with a percentage of fines below 5%. Considerations con-cerning the allowable settlement of the building required a stratum of about 6.0 m thickness below the structure to have an average relative density of $D_r = 75\%$, corresponding to the existing circumstances with a static CPT resistance value of 20 MN/m². CPT values of the natural ground before any compaction were generally below 12 MN/m².

Prior experience in Berlin sand with the depth vibrator, which was cho-sen for the project, indicated a vibro probe spacing between 2.0 and 3.0 m for a triangular compaction pattern. The test field arrangements as shown in Figure 3.50 was designed according to Figure 3.17 for three potential probe spacings of 2.0, 2.5, and 3.0 m. Static CPT results before and after vibro compaction are given in Figure 3.51. Since the specified cone resis-tance of 20 MN/m² was not achieved with the 3.0 m grid in the depth range of 5.0–9.0 m below surface a grid spacing of 2.5 m was finally proposed for the vibro compaction works.

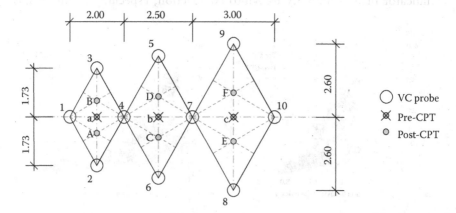

Figure 3.50 Test field to establish vibro compaction probe spacing. VC, vibro compaction. (Courtesy of GuD, Berlin, Germany.)

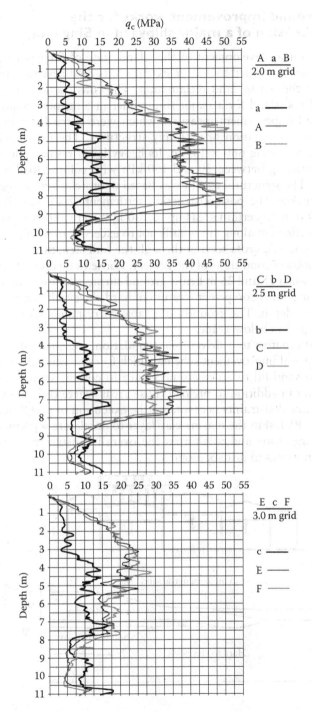

Figure 3.51 Pre- and post-CPT results. (Courtesy of GuD, Berlin, Germany.)

3.6.7 Ground improvement works for the extension of a major shipyard in Singapore

The economical development of the island state of Singapore during the past 50 years depended to a large extent on its readiness to extend its territory into the sea by land reclamation. The construction of a large hull shop on Tuas Island for a major shipbuilder in Singapore required extensive ground improvement measures which were carried out in 2013 and 2014 for the foundation of its heavy structures. The main facility is the hull shop measuring 680 × 185 m. Its heavy loads result in floor loadings typically ranging between 20 and 160 kPa with column loadings of 100–2000 kN. The structural design was based upon relatively stringent deformation criteria: The postconstruction settlement was generally restricted to only 50 mm (in certain areas up to 100 mm were allowed) with the differential settlement along crane rails not to exceed 1 in 1000.

Figure 3.52 is a cross section through the hull structure with the representative subsoil conditions of the project. Loose to medium dense sand fill ranges between 15 and 26 m below ground surface, followed by ca. 10 m of stiff marine clay. A competent stratum of hard clayey silt is generally found below 26 m depth. The groundwater table corresponds with the sea level and is about 4 m below grade.

As an alternative to a driven pile foundation, the successful contractor proposed a soil improvement scheme which foresaw generally the compaction of the sand fill to a density necessary to comply with the settlement criteria, and in addition to preload those areas where the marine clay was present. Time constraints necessitated the execution of prefabricated vertical drains (PVDs) in the marine clay layer to accelerate settlements during the surcharge time and to shorten the overall construction time for the foundation works to just six months.

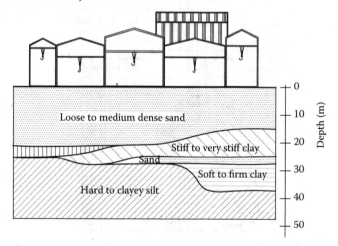

Figure 3.52 Representative soil conditions. (Courtesy of Keller Group plc, London, UK.)

In this period, an area of 126,000 m² was treated by vibro compaction to depths between 11 and 26 m. In areas where preloading was necessary (ca. 30,000 m²), the vibro compaction treatment depth increased by 5.6 m, the height of the sand surcharge. PVDs were carried out beforehand to depths of up to 40 m (ca. 20 m sand fill, 10 m sandy clay fill, and 10 m marine clay) to reach the bottom of the marine clay layer. Drain distances were designed conservatively to 1 m, in square grid, in order to achieve 80% consolidation within three months' time under the 5.6 m sand surcharge. Consequently, some 1,300,000 lin. m of prefabricated vertical drains were executed.

To accommodate the deformation criteria for the various parts of the large hull structure, the postvibro compaction soil stiffness was specified by the soil consultant in terms of carefully selected, standardized CPT acceptance criteria. Several different criteria were proposed according to the design loading and allowable settlements, but generally varied between 8 and 16 MPa, in a stepwise manner increasing with depth.

The sand fill material was hydraulically deposited by a standard dredging operation. As in many instances in Singapore, it is taken from the seabed and imported to the reclamation area. It consists primarily of quartz sand, with less than 10% carbonate content, and can be described by the USCS as SW material (well graded clean sand with less than 5% fines) and reaches after dredging generally just medium density.

The specialist contractor chose to perform precontract compaction trials to establish the necessary compaction parameters (probe distance, compaction time) to meet these specifications. Figure 3.53 shows the locations of the five CPTs performed within a 3.6 m triangular compaction pattern.

With the five postcompaction CPT results, an area weighted average was calculated which then served as reference curve for a treated representative

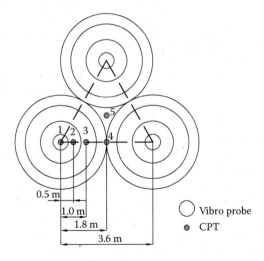

Figure 3.53 Post-CPT locations to establish acceptance criteria for vibro compaction.

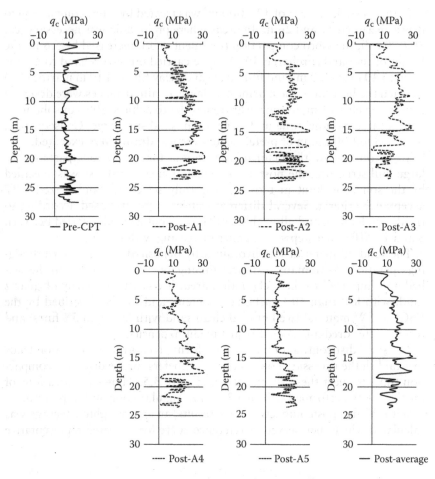

Figure 3.54 Pre- and postcompaction CPT results for trial compaction. (Courtesy of Keller Group plc, London, UK.)

sand volume to be compared for acceptance with the specified target curves. Figure 3.54 shows the precompaction CPT curve together with the five post-compaction curves and the calculated average CPT curve (post-average) for one of the trial patterns.

The vibro compaction work was performed with S700 Keller depth vibrators (see Table 3.1) in three different triangular patterns depending on the target specifications. In total about 400,000 lin. m of vibro compaction was performed on an area of 126,000 m². The compaction areas were divided into panels of 3000 m². Eight quality control tests were performed in each panel. In general, the sand fill material responded rather well to the vibratory treatment which was visible by a surface settlement, equivalent to about 5% of the compaction depth, which was compensated during execution of the works by imported sand fill.

Chapter 4

Improvement of fine-grained and cohesive soils by vibro replacement stone columns

4.1 VIBRO REPLACEMENT STONE COLUMN TECHNIQUE

We have seen from the description of the development of the deep vibratory processes that vibro stone columns extend the limits of application of these ground improvement methods from predominantly noncohesive into fine-grained cohesive soils. With regard to the grain size distribution of the soils to be improved, vibro replacement stone columns are used when conditions prevail as represented by areas outside of zone B in Figure 3.13. In the following, it will be shown that the load-carrying capability of soils so improved depends primarily from the interaction of the stone columns with the surrounding soil. The columns are supported by the soil which they replace and which requires therefore a certain minimum strength.

With increasing fines content in the grain size distribution of sands, the spacing between compaction probes needs to be decreased to obtain sufficient compaction when using the vibro compaction technique, as discussed in Chapter 3. At the same time, the addition of coarse backfill material helps to achieve the desired density. However, when the fines content exceeds about 10%, the vibrations emanating from the depth vibrator can no longer separate the soil particles from each other to achieve a closer density, owing to their cohesive forces. Improvement of such soils, which are generally relatively soft and impervious, can be effected by introducing stone columns. Two methods are currently being distinguished by the strength range of the soils to be improved, generally in terms of their undrained shear strength c_u.

The *vibro replacement wet method* is employed only in water-bearing soft soils within a strength range of around $c_u = 10–30$ kN/m². In these soils the depth vibrator normally sinks by its own weight, and that of the extension tubes, to the desired depth, helped only by the low pressure, large volume bottom jets. Collapsing soil from the side walls of the hole is transported in the water flow to the surface where an effective sludge and water management has to take care of it. To avoid pollution, only water that has been largely freed from soil particles is allowed to be disposed of into available recipients. The water can contain significant amounts of suspended silt and clay which require proper handling in settling ponds and standing pools before the water can be reused for the stone column production or before it is released. These procedures need to be well organized to avoid disruption of the work or slowing down of production. In this context it is, however, necessary to remember that sufficient water volume is an integral part of the wet vibro replacement stone column method to guarantee stability of the hole and its diameter.

On reaching the desired depth, the vibrator is often completely withdrawn from the bore before it is allowed to repenetrate rapidly—sometimes a few times—to full depth, in this way allowing the bore to be cleaned from loosened soil material and its diameter to increase. In most cases the bore created is at least temporarily stable, and coarse backfill can now be filled, preferably in small doses, into the bore hole. The vibrator is then lowered again to full depth and the vibrations, together with a slight upward and downward motion with amplitude of generally not more than a meter, cause the backfill to be compacted and rammed into the sides of the bore. With increasing density of the backfill, vibrations are also transmitted to the surrounding soil. The resulting shear stresses may cause the soil to collapse leading to a further increased bore diameter as the continuing water flow, movement of the vibrator and of the backfill, transports this fine material to the surface. When equilibrium is reached, column building begins as further stone is added as described or through the annulus around the vibrator remaining in the bore hole. Rising resistance indicated by slowing down the sinking rate of the machine, accompanied by increased power consumption of the vibrator motor, is a sign that column building is completed at this level and that repetition of this procedure should start at the next higher level.

In this manner, an stone column is formed up to ground surface in a self-compensating way with diameters of about 0.8–1.2 m, depending on the soil resistance and the shearing and flushing action over the time of the build-up (Figure 4.1a).

It is evident that this replacement process causes only relatively little disturbance to the surrounding native soil, without apparent smear damage (Greenwood, 1976), allowing, under loading conditions, the pore water to freely drain into the columns, resulting in a considerably accelerated consolidation process (see Section 4.3.4).

In more stable insensitive cohesive soils with strength values $c_u = 30$–50 kN/m^2, the *dry vibro displacement method* is applied whereby the depth vibrator penetrates by vibratory impact and by its own weight, sometimes increased by that of the heavy extension tubes, and always helped by compressed air emanating through the bottom jets of the machine. When the design depth is reached, it is always necessary to extract the vibrator completely from the ground to allow coarse backfill material to be introduced in small quantities into the bore. The compressed air is used primarily to prevent

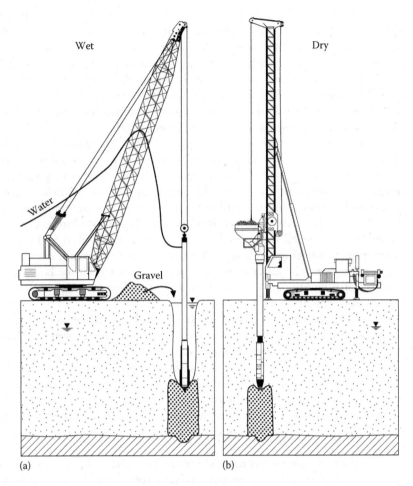

Figure 4.1 Stone columns built (a) by the wet vibro replacement method and (b) by the dry displacement bottom feed method.

the bore from collapsing as a result of the suction developing when the vibrator is withdrawn, as it is generally in a tight fit in the ground. The column is built by tipping the backfill into the bore hole and by lowering the machine back into it in lifts of about half a meter. By the action of the depth vibrator, the coarse backfill is compacted and displaced laterally and downward forming in this way a tightly compacted stone column of about 0.5–0.6 m diameter which is well interlocked with the surrounding sheared soil. To safeguard column verticality and to assist penetration into stiff soil, custom-built cranes with leaders were developed, replacing the standard cranes as base machines.

As this method has its downsides with deeper penetration, it being relatively slow and requiring well-supervised workmanship, a special depth vibrator was developed avoiding the need to remove it from the ground for material backfilling and at the same time ensuring column continuity for deeper columns in much softer soils ($c_u = 10$ kN/m²). With this *dry displacement bottom feed method* the backfill is discharged directly at the point of the vibrator through a pipe that is fitted to its outside and which is fed by a backfill container with an airlock situated on top of the extension tubes. The system is best used with dedicated cranes for the depth of less than 20 m (these are called vibrocats and develop considerable pull-down forces to compensate the increased friction force developing as a result of the increased depth vibrator cross section) (Section 4.2). For greater depths standard crawler cranes can be used. A continuous column is formed by pulling the vibrator in steps of 0.5–1.0 m within the ground, in this way allowing the backfill to flow into the bore hole, helped by a moderate flow of compressed air. By repenetrating until resistance is met, the coarse material is compacted and laterally displaced into the soil. This procedure is repeated leaving behind a well-compacted stone column of about 0.5–0.8 m in diameter up to working grade level. The method is to a certain degree self-compensating since in softer soils column diameters tend to be larger than in stiffer soils. Figure 4.1b shows a typical plant while forming vibro stone columns with the dry bottom feed method.

It is evident that the wet system primarily replaces native soil by coarse backfill with relatively little lateral displacement; the dry system is, in contrast, in the first place a displacement method. Nevertheless both methods are today generally named vibro replacement stone column methods, but we will see later that displacement of the surrounding soil can add extra strength to columns of otherwise equal dimensions.

The grading of backfill stone for the vibro stone columns differs slightly between the different methods of construction.

- For the replacement wet method rounded or subangular stone or gravel 30–60 mm in size and comparatively uniformly graded is used which passes easily through the annulus around the machine. It is always found, on excavation of a stone column, that the voids between the stones, which are in close contact with each other, are always found to be filled with the coarser particles (sand and coarse silt) from the native soil ensuring a well compacted stone column. The finer parts of the native soil are transported to the ground surface with the flushing water.
- For the bottom feed method, finer gravel or crushed stone of 10–40 mm is necessary to pass through the delivery system including the feed pipe to the vibrator point. Well-graded sand can also be used should no coarser backfill be available, but this will forfeit a considerable amount of column strength in comparison to the coarser stone backfill (Section 4.3).

The material of the stone backfill should be environmentally acceptable and can be taken either from natural gravel deposits, crushed natural stone, or from recycled concrete or brick debris. The material itself should be of sufficient hardness and strength to resist the strong abrasive forces resulting from the vibratory action of the depth vibrator. It should not contain any organic and other deleterious material. It should be chemically inert and should resist—in cases of aggressive groundwater conditions—any potential ion attack during the lifetime of the foundations.

Regardless of the method, whether suspension from crawler cranes, or guided by leaders, vibrators are always applied in the vertical orientation. The vibrator forms a vertical hole largely by displacement, sometimes enlarged by water flushing, ultimately hanging as a pendulum in the bore hole. While crane-hung vibrators receive no guidance when penetrating and during column construction, the special rigs (Section 4.2) are generally equipped with vertical leaders providing excellent guidance to the vibrator in all working phases. Varying soil strengths during penetration, and obstructions, may cause the vibrator to depart from its vertical position. It is very important that these deviations are detected and corrected in good time in order to ensure the verticality of the stiffening columns.

Vibro stone columns improve the foundation ground because they are stiffer and of higher strength than the soil which they have replaced. The reinforcing effect depends primarily on the column diameter and the distance between adjacent columns. They are generally installed with center spacing of 1.5–3.5 m, in rows below strip footings and in triangular or square patterns below single footings and widespread loaded areas. As with vibro compaction, the upper 0.3–0.5 m of a stone column are generally less well compacted as a result of vibrator shape and missing lateral support from the surrounding soil. Therefore, before concreting work begins, the foundation area is excavated to the required level, about 0.3 m of well-graded coarse gravel is placed on top of the treated area and the whole surface prepared in this way is well compacted with a standard surface compactor.

Production rates depend, in the main, on the stiffness of the ground needing improvement but also on the stone volume consumed for the stone columns, on the characteristics of the vibrator, and on the efficiency of the working crew. Rates generally range between 150 and 450 m of stone columns for a 10-h shift per vibrator unit.

4.2 SPECIAL EQUIPMENT

For the construction of vibro replacement stone columns employing the wet method, essentially the same equipment is needed as for the vibro compaction process and as described in Sections 3.1 and 3.2. When working in cohesive soils, the water management and sludge handling on-site requires special attention. Silt and clay particles in suspension cannot be completely or easily removed from the water before it is released into natural or artificial recipients. If this shortcoming is environmentally acceptable, the construction of stone columns with diameters in excess of 0.6 m can best be achieved by employing vibrators with larger diameters (see Table 3.1) and high-volume, low-pressure water pumps. Very often when soil conditions prevail where layers of sand alternate with cohesive soils, the use of high-amplitude, low-frequency vibrators is advisable as these machines allow both efficient compaction of the sand layer and forming of stone columns in one operation.

With the dry vibro displacement method, vibrators with higher frequencies and smaller diameters are used which generally penetrate more effectively into the ground. The water pipes are replaced by smaller diameter air pipes also leading to the vibrator point. Compressed air generated by standard compressors (9 m³/min, 5 bar), often mounted together with the electric generator or hydraulic power pack on the rear of the crawler crane, produces a moderate airflow, just strong enough to avoid suction developing when the vibrator is withdrawn from the ground, and sufficient to keep the stone backfill flowing in the stone supply system when the bottom feed method is used. Strong airflow at high pressure should be avoided as it often does more damage to the fabric of soft soils than helping to remove the vibrator from the bore hole (Greenwood and Kirsch, 1984). It can also cause unwanted heave of the ground.

Today, we mainly use purpose-built machines with vertical leader and on-board generator and compressor. Since these machines are generally mounted on crawlers they are often called vibrocats. A special gravel container at its front allows the stone backfill to be funneled into the bore hole, generally in small quantities just enough to form about 1 m of stone column length. As effective penetration aids, these machines develop a pull-down force of up to about 300 kN. When obstructions or stiff layers prevent the vibrator from reaching the design depth, a more powerful vibrator can resolve the problem or preboring with continuous flight augers might become necessary, preventing uneconomical slow production and unnecessary wear and tear on the depth vibrator.

The special bottom feed equipment both for crane-hung and vibrocat operation is today the standard setup for dry stone column construction. Figure 4.2 shows a vibrocat rig with a special depth vibrator whose extension tubes have been transformed into gravel containers, feeding the stone backfill through a half-moon-shaped gravel pipe to the point of the depth vibrator.

The construction details of a bottom feed vibrator are shown in Figure 4.3 in a schematic mode. Since its cross section is substantially increased as a result of the attached gravel pipes, penetration resistance of the soil increases considerably, requiring much more pull-down capability of the vibrocat—hence the need for the strong activating forces available in modern machines. As can also be seen from Figure 4.3, the cross section through the vibrator and gravel pipe is of oval shape, no longer circular. The stone columns so constructed are also of an oval shape and need to be

Figure 4.2 Vibrocat with bottom feed vibrator. (Courtesy of Keller Group plc, London, UK.)

transformed into circular columns of equal cross-sectional area in relevant design calculations.

As already mentioned, the vibrocat carries a generator and a compressor and is equipped with a gravel hopper which is filled with stone backfill at ground level and which slides up the vertical leader to the top of the extension/gravel container pipe. The hopper is emptied via an airlock into the extension tube acting as gravel container with a capacity of approximately 0.1 m³/m of length, allowing the stone column construction to continue, usually without any interruption. With modern machines this process can be carried out in an automated way, whereby the vibrator lifts, including stone filling, repenetration and compaction are predetermined at the beginning of the site and repeated in the same manner for all stone columns of any given project. Feeding of the stone container is equally automated, to allow the operator to concentrate on the process controls displayed in front of him in the cabin of the rig. The need to refill the gravel container is indicated to the operator via a level control inside the gravel container. All process data as displayed to the operator are stored as

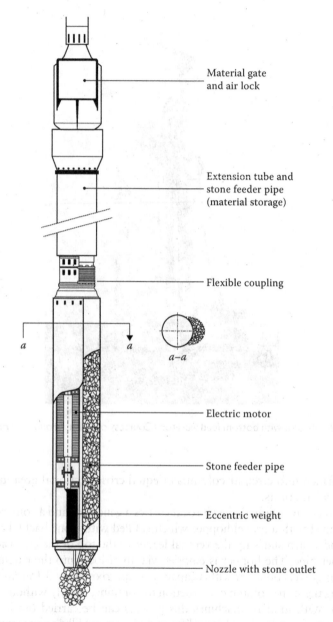

Material gate and air lock

Extension tube and stone feeder pipe (material storage)

Flexible coupling

a—a

Electric motor

Stone feeder pipe

Eccentric weight

Nozzle with stone outlet

Figure 4.3 Special features of a bottom feed vibrator.

a function of time and can be plotted for each individual stone column in real time or at the end of the shift. These data include for each stone column: time of execution, depth, power, and stone consumption over depth, total stone consumption, verticality of the leader, and optionally also the pressure of the airflow and the stone column diameter.

Modern data acquisition systems controlling both column building and operational parameters of the equipment employed will eventually, together with GPS systems, allow the contractor to construct large number of stone columns per working hour of uniform quality and in a fully automated way. Telemetric systems are already used to transmit not only service and machine performance indicators (fuel consumption, oil pressure, etc.) but also machine location, and all relevant process data as compiled by the control and data acquisition unit employed.

To limit or even avoid operational errors and mistakes when constructing these stone columns a specialist contractor recently proposed an operator guidance system (Betterground, 2014) enabling also less experienced operators to install the stone columns according to specifications in difficult soils. Based on the data acquired during the installation process, the onboard computer helps the operator to decide when the stone filling process for a predetermined column diameter in specific soils is completed and the vibrator can be lifted into the next position where the filling process will start again. The system allows programming the up and down movements of the vibrator, for example, in the following way (Betterground, 2014):

- Start column at a certain depth
- Build at least an 80 cm column, unless a specific amperage (say 240 A) of the depth vibrator motor is reached in which case also smaller column diameters are allowed
- If the amperage stays below a certain value (say 150 A) soft soil is encountered and a larger column diameter (say 140 cm) is to be built or the threshold amperage is reached, whatever comes first

Needless to say that the interpretation of the real-time recordings of trial vibro stone columns carried out at the beginning of a foundation project requires considerable experience.

Modern vibrocats are designed to install stone columns to depths of up to 20 m. To expedite setting up of the vibro replacement equipment on-site special rigs were developed for shorter, more frequently used columns of up to 10 m, which represents the depth range of the majority of sites, allowing the complete bottom feed depth vibrator to remain on the rig when being transported. To make use of standard excavators (which are plentifully available on the construction equipment market) as base machines for carrying depth vibrators to construct vibro stone columns specialist contractors have developed purposely built attachments. These provide essentially the same characteristics necessary for safe and controllable execution of stone columns; they do however, develop only a moderate pull-down force. Figure 4.4 shows such equipment capable of building 11 m deep stone columns.

For stone columns deeper than 20 m, and for columns to be constructed over open waters, the crane-hung bottom feed method can be used. Base machines for this purpose are generally heavy crawler cranes that carry the

Figure 4.4 Attachment for standard excavators to carry out stone column installations. (Courtesy of Keller Group plc, London, UK.)

depth vibrator with standard extension tubes and welded on gravel pipes. A gravel container capable of carrying about 2 m³ of stone backfill is situated on top of the extension tube. Gravel is filled at ground level by payloaders into a hopper, which is pulled upward and delivers its contents via a receiving mechanism into the gravel container, from where it is released by special controls into the gravel pipe. Figure 4.5 shows a crane-hung bottom feed assembly for constructing deep stone columns.

When stone columns need to be built through deep water, supply of the backfill stone to the machine is often rather difficult to achieve. For such applications, special pneumatic or hydraulic transport systems for the gravel were developed which replace mechanical stone delivery by hoppers. Figure 4.6 shows such an arrangement by which the gravel is transported hydraulically via special pumps and hoses into the receiving tank from where stone supply takes place through the gravel pipe in an uninterrupted manner under constant air pressure until it is released at the vibrator nose cone into the soil. Up and down movements compact the stone and displace it into the soil, thus forming a stone column up to the sea bottom or ground level.

In very deep water, the length of the vibrator unit with its extension tubes and the associated stone storage and delivery system may become excessively long, requiring very large cranes to handle the whole apparatus. To avoid this drawback, specialist contractors have developed recently

Figure 4.5 Crane-hung dry bottom feed system for deep stone columns (S-Alpha system). (Courtesy of Keller Group plc, London, UK.)

submersible stone storage systems allowing the usage of moderate size cranes for the construction of stone columns to great depths without compromising their quality and execution accuracy.

4.3 PRINCIPAL BEHAVIOR OF VIBRO STONE COLUMNS UNDER LOAD AND THEIR DESIGN

4.3.1 Overview and definitions

Vibro stone columns are in general used to improve characteristics such as the strength and compressibility of fine-grained cohesive soils that are deemed insufficient for the geotechnical purpose. These soils are predominantly water-bearing silty and clayey sands, sandy silts, and silt and clay mixtures. Their main characteristics, water content, consistency, and shear strength, are shown in Figure 4.7. These soil properties, together with those of the stone columns, influence the behavior of the composite soil–stone column matrix.

Certain definitions and some of the terminology used in conjunction with ground improvement by vibro replacement stone columns are further

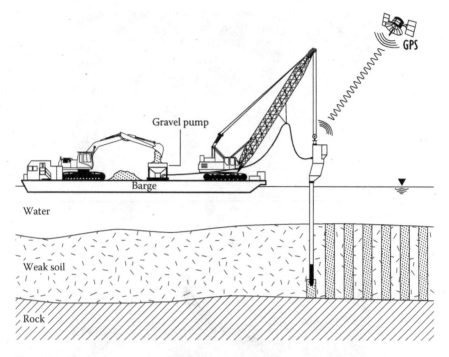

Figure 4.6 Crane-hung bottom feed system with gravel pump for stone supply.

discussed in this section. The terms *vibro replacement stone column, vibro stone column*, and *stone column* are used synonymously. In contrast to a single stone column, a group of stone columns is represented by a certain number of stone columns that are arranged in such a way that they influence each other in the ground when loaded and this is described as the *group effect*. In contrast to the isolated stone column and the column group, large numbers of regularly arranged stone columns of equal diameter and equal separating distance are normally referred to as an *infinite grid pattern of stone columns*. Their behavior under load can be described by the unit cell concept, whereby—irrespective of the position of the column within the grid—all columns behave equally, and the description of the mechanical model of a single column is representative for the whole grid. Figure 4.8 gives the geometry of the unit cell for triangular and square grid patterns.

The equivalent diameter of the circular unit cell d_e as a function of the column distance b can be calculated as

$$d_e = C \cdot b \tag{4.1}$$

with

    $C = 1.05$ for the triangular pattern
    $C = 1.13$ for the square pattern

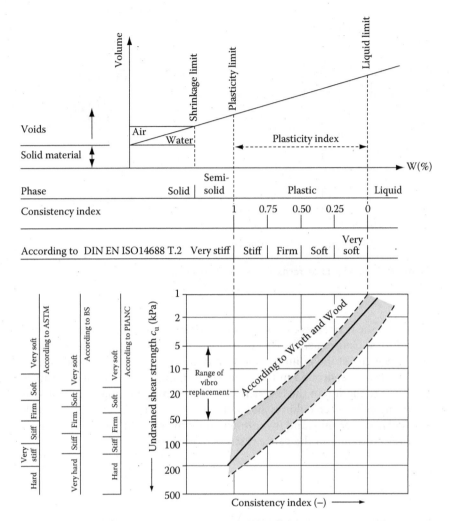

Figure 4.7 Water content, consistency, and shear strength of fine-grained cohesive soils. (Based on CUR, *Building on Soft Soils*. Centre for Civil Engineering Research and Codes. A.A. Balkema, Rotterdam, the Netherlands, 1996.)

Another important parameter of the infinite grid pattern is the area replacement ratio a_c calculated as ratio of column area A_c and total unit cell area A:

$$a_c = \frac{A_c}{A} = \frac{1}{C^2} \cdot \left(\frac{d}{d_e} \right)^2 \tag{4.2}$$

As the unit cell is symmetric by geometry, a uniform load evenly distributed over the infinitely extended area and applied by a rigid raft will not

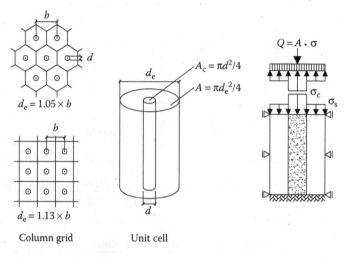

Figure 4.8 Column grid patterns and unit cell concept.

cause any shear stresses or horizontal deformations to develop at its outside boundaries. Consequently, the uniform loading will remain within the unit cell. Since the stone column is stiffer than the surrounding soil, vertical stresses are concentrated in the stone column material with an accompanying stress reduction in the less stiff soil surrounding the column. This stress distribution between column and soil within the unit cell is generally called stress concentration n and is represented by the ratio of the vertical stresses σ_c in the column and the vertical stress σ_s in the soil:

$$n = \frac{\sigma_c}{\sigma_s} \tag{4.3}$$

Equilibrium with the acting vertical stress σ is given by the relationship

$$\sigma = \sigma_c \frac{A_c}{A} + \sigma_s \left(1 - \frac{A_c}{A}\right) = a_c \cdot \sigma_c + (1 - a_c) \cdot \sigma_s \tag{4.4}$$

which leads to expressions for σ_c and σ_s as follows:

$$\sigma_c = \sigma \cdot \frac{n}{(1 + (n-1) \cdot a_c)} = n_c \cdot \sigma \tag{4.5}$$

with

$$n_c = \frac{n}{(1 + (n-1) \cdot a_c)} \tag{4.6}$$

$$\sigma_s = \sigma \cdot \frac{1}{(1 + (n-1) \cdot a_c)} = n_s \cdot \sigma \tag{4.7}$$

with

$$n_s = \frac{1}{(1 + (n-1) \cdot a_c)} \tag{4.8}$$

The expressions n_c and n_s represent the ratio of stresses in the stone column and the soil, respectively, to the average stress σ acting on the unit cell and are connected with each other and the stress concentration factor n by the expression:

$$n_c = n \cdot n_s \tag{4.9}$$

Equation 4.7 can be rewritten for the ratio of the acting stress σ with the soil stress σ_s:

$$\frac{\sigma}{\sigma_s} = 1 + (n-1) \cdot a_c \tag{4.10}$$

The unit cell concept stipulates equal settlements for stone column and tributary soil:

$$s_s = s_c \tag{4.11}$$

Priebe (1976) was the first to define the settlement improvement β as the ratio of the settlement s of the untreated soil and the settlement s_i of the improved soil. Using the assumption that settlements behave directly proportionally with their stresses, based on Equation 4.11, an expression for the settlement improvement factor β can be derived from the unit cell concept:

$$\beta = \frac{s}{s_i} = \frac{\sigma}{\sigma_s} = 1 + (n-1) \cdot a_c \tag{4.12}$$

In reality, where these ideal conditions do not prevail, the stress concentration factor n depends not only on variables such as the area replacement ratio $a_c = A_c/A$, but also on the length of the stone column and particularly on the relative stiffness of column and soil. Values of n measured in the field are about 1.5–5.0, usually taken close to foundation level. As Barksdale and Bachus (1983) have already pointed out, n will increase with consolidation time and will not decrease below its value at the end of the primary settlement.

The expressions according to Equations 4.6 and 4.8 are helpful in both settlement calculations and stability analyses for cases where the unit cell concept can be applied and where the stress concentration n can be estimated with sufficient accuracy from Equation 4.12. With decreasing numbers of stone columns in large groups the accuracy of the method decreases.

The main effect of a ground improvement measure is settlement reduction expressed by the improvement factor β as the ratio of the settlement without stone columns and the settlement after ground improvement. The different behavior of a stone column within a group is governed by its relative position within it and influencing its overall performance. With increasing numbers of columns, group behavior approaches that of the infinite column grid. It is difficult to define the number of stone columns that no longer behave as a column group since other factors such as improvement depth, size, and rigidity of the foundation have an influence. For practical reasons, more than 50 stone columns regularly arranged below a foundation slab may be addressed as an infinite pattern of columns when the ratio of foundation width to column depth is at least 3.

The effect of stone columns on the ground to be improved is manifold. The higher stiffness of the column material in relation to that of the soil leads to load concentrations on the columns, thereby reducing the settlement. Since the column material is considerably more permeable than the soil, stone columns act as drains when constructed and loaded in water-bearing soils, reducing consolidation time substantially. The fill material also has a much higher shear strength compared to the original soil; therefore, also increasing its bearing capacity. The improvement effect is increased by the load concentration on the columns also leading to increased stability of structures when supported by vibro stone columns. In saturated soils, the existence of stone columns reduces the liquefaction potential of silt and sand deposits during dynamic loading or earthquakes. The combination of densification, drainage, and increased shear strength prevents total loss of strength during a seismic event. To summarize, the main effects of the vibro replacement method are

- Reduction of settlements.
- Reduction of consolidation time.
- Increase of bearing capacity.
- Reduction of liquefaction potential.

In parallel to the large variety of applications of vibro stone columns, and as a result of the technical development and better understanding of ground improvement in general, various computational and design methods have been proposed. These can be distinguished from each other by the computation approach, by the column geometry that is being considered, and by the design objective, which could be bearing capacity, settlement reduction or acceleration, and earthquake risk mitigation. Some of these methods are exclusively based on empirical findings, while others use analytical approaches of the cylindrical

Table 4.1 Necessary design parameters for commonly used computation methods

	Priebe	Goughnour/ Bayuk	Hughes/ Withers	Brauns	Van Impe/ Madhav	Balaam/ Booker
Column						
Unit weight	•	•				•
Friction angle	•	•	•	•	•	•
Stiffness	•	•			•	•
Poisson's ratio	•	•			•	•
Angle of dilatancy					•	•
Soil						
Unit weight	•	•				•
Friction angle		•				
Cohesion						
Undrained shear strength			•	•		
Stiffness	•				•	•
Poisson's ratio	•				•	•
Initial stress state		•				
Loading		•				

half space or introduce empirical parameters into their design concept. A more recent group of methods develops design diagrams that are based on numerical calculations.

Most design methods deal with the isolated column or the infinite grid of vibro stone columns. Almost all methods apply the principle of the circular unit cell (Figure 4.8). Only a few methods consider the behavior of column groups.

Settlement reduction is of particular practical importance for the design of vibro stone columns, where the allowable bearing pressure needs to be secured in a separate step by also determining the ultimate bearing capacity of the system. In addition, consolidation time or the reduction of the liquefaction potential may also have to be determined.

A selection of the commonly used design methods can be found in Table 4.1 together with their key design targets. The table also gives the parameters necessary for the design calculation. These methods are empirical and analytical, purely analytical, or analytical and numerical.

Ideally, the design models should reflect the interactions between load application, stone columns, and surrounding soil as will be discussed in the Sections 4.6.1 through 4.6.3. In particular, the elastic–plastic behavior and the dilatancy of the stone column material are important features for the model, and the model is not only valid for the two special cases of the isolated column and for the infinite column grid, but it also governs the column group behavior (Section 4.6.4). However, such high demands are not yet state of the art in designing stone columns, since all methods are still based on far-reaching simplifying assumptions (Soyez, 1987; Bergado et al., 1994; Kirsch, 2004).

4.3.2 Load-carrying mechanism and settlement estimation

Principal loading situations for stone column foundations are shown in Figure 4.9. Foundation stresses acting on the improved soil lead to a stress concentration in the material of higher stiffness, that is, in the stone columns, which results in a stress relief of the soft surrounding soil. From the assumption that vertical deformations of stone column and soil are equal in horizontal planes, with the reduced load on the native soil, follows the settlement reduction of this ground improvement system. Various researchers have confirmed this assumption in the laboratory (Nahrgang, 1976),

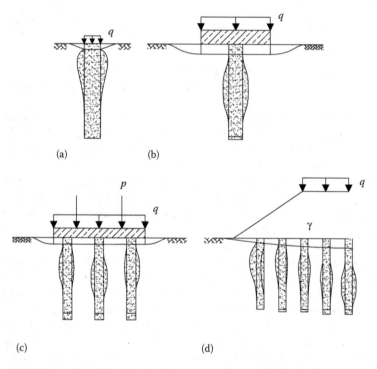

Figure 4.9 Different loading scenarios for stone columns: (a) Single column, (b) Single column below footing, (c) Column group below footing, and (d) Column grid below embankment.

by numerical calculations (Balaam and Poulos, 1983) and by in-situ measurements (Gruber, 1994; Kirsch, 2004).

In contrast to the load-carrying mechanism of a rigid foundation element, such as a pile, where the load transfer into the ground is by toe resistance and friction at the pile perimeter, stone columns transmit their load into the soil by stimulating its horizontal earth pressure without relative displacement between column and soil. In this state of contained compression, the stone column, which is characterized by high density and stiffness, will ultimately fail by bulging as a result of the high column load and the minimal supporting capacity of the surrounding soil. Owing to its high density, the column yields locally in its upper part under the shearing forces by lateral bulging, counteracted by the surrounding soil, depending on its stiffness and strength. This horizontal deformation is further increased by the dilatancy of the column material. This volume increase was measured to about 9%, albeit that there is considerable axial column compression (Kirsch, 2004).

The hypothesis that a column is triaxially loaded in confined compression leading to the peculiar horizontal deformation, as described, has been verified by numerous researchers (Greenwood, 1970; Hughes and Withers, 1974; Hughes et al., 1975; Brauns, 1980; Barksdale and Bachus, 1983; Kirsch, 2004). This has been done theoretically, in model tests and in field tests on excavated columns. Figure 4.10 shows such a model test to investigate column group behavior. The different behavior of a center column in comparison to one at the edge is clearly visible.

Figure 4.10 Model tests on four and five stone columns to study column group behavior.

As column bulging is followed by lateral support, triggering soil deformation which in response further increases the horizontal stresses, the overall load-carrying mechanism is controlled by the interaction of the load application with columns and native soil. Figure 4.11 shows the different features of this interaction of the composite of column and soil with the load application.

In most applications of the vibro replacement method, the column heads are not in direct contact with the applied load, which is almost always transferred via a load distribution layer. Occasionally, horizontal load-carrying layers consisting of a combination of coarse granular material and special geotextiles are used, particularly below embankments and similar structures, where horizontal forces need to be controlled. These load-transfer layers can provide an additional load concentration at the column head when they are thick enough in relation to column diameter and distance for an arching effect to develop. This is beneficial for the design of the foundation slab, but the load distribution has no significant effect on the overall settlement (Figure 4.12).

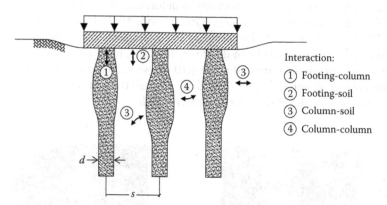

Figure 4.11 Interaction of load application with stone columns and soil.

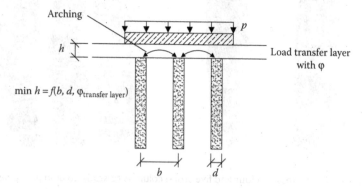

Figure 4.12 Load-transfer layer and arching effect.

Almost all approaches to calculating the settlement reduction by the use of vibro stone columns are based on the infinite column grid. Few deal with the quantitative behavior of column groups, and all make major simplifying assumptions. Investigations of the settlement behavior of isolated columns are rare and of little practical use. Frequently, however, single-column load testing is performed as a quality control measure and to predict settlement performance of structures, a method requiring particularly careful evaluation (see Section 4.4).

Greenwood (1970) was the first to propose a design chart for the infinite column grid providing a relationship for the reciprocal improvement factor $1/\beta$ as a function of the column distance in the grid for cohesive soils with strengths between $c_u = 20$ and 40 kPa (Figure 4.13). From the context of the paper the column diameter can be assumed as 0.9–1.0 m for wet and as 0.6–0.7 m for dry column construction.

Of the many design methods, only two are of practical importance today— the Priebe method and the Goughnour and Bayuk method—which will be further discussed in this section. The Priebe method is widely used in Europe and was presented for the first time in 1976, with improvements and alterations made to it later (Priebe, 1976, 1988, 1995, 2003). The method uses the unit cell concept as shown in Figure 4.8 describing the stress situation of an infinite grid pattern of stone columns loaded with the vertical stress σ via a rigid foundation raft. The deformation of the stone column is approximated by the cylindrical cavity expansion method, which was proposed by Gibson and Anderson (1961), for describing the pressure meter deformations.

From equal settlements for stone column and tributary soil within the unit cell, and equilibrium between foundation load and the loading shares

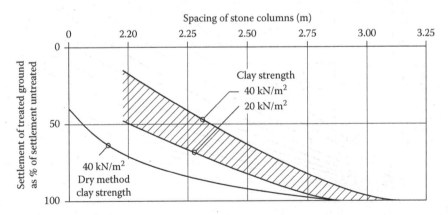

Figure 4.13 Settlement diagram for stone columns in soft uniform clay. Note: Curves neglect immediate settlement and shear displacement. Columns assumed resting on firm clay, sand or harder ground. (After Greenwood, D.A., Mechanical improvement of soils below ground surface, in *Ground Engineering*, The Institution of Civil Engineers, London, UK, 1970.)

of column and soil, comes an expression for the improvement factor β as the ratio of the settlement of the untreated (s) and the improved ground (s_i):

$$\beta = \frac{s}{s_i} = 1 + \frac{A_c}{A} \cdot \left[\frac{1/2 + f(\mu_s, A_c/A)}{K_{ac} \cdot f(\mu_s, A_c/A)} - 1 \right] \tag{4.13}$$

with

$$f(\mu_s, A_c/A) = \frac{1 - \mu_s^2}{1 - \mu_s - 2\mu_s^2} \cdot \frac{(1 - 2\mu_s) \cdot (1 - A_c/A)}{1 - 2\mu_s + A_c/A} \tag{4.14}$$

and

$$K_{ac} = \tan^2(45° - \varphi_c/2) \tag{4.15}$$

In these equations the notations of the unit cell from Figure 4.8 are used; with φ_c representing the friction angle of the column material and μ_s the Poisson's ratio of the soil.

For easy use of the method, Priebe developed a diagram for the improvement factor β for the infinite column grid as a function of the ratio A/A_c (total grid area A to column area A_c) with friction angle φ_c of the column material as sole parameter. In Figure 4.14, this diagram is presented with the area replacement ratio $a_c = A_c/A$ as abscissa for the practically relevant range of a_c between 0.05 and 0.35. The graph was also extended to column material friction angles in excess of 45°.

In practice, the method requires a conventional settlement calculation to be carried out for the untreated ground, usually by the summation of

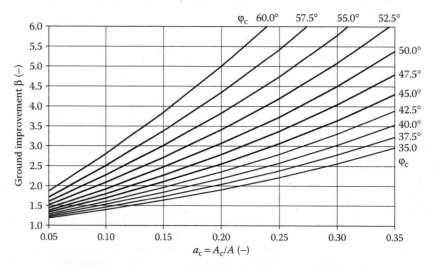

Figure 4.14 Basic ground improvement design chart for infinite column grid patterns with a_c between 0.05 and 0.35 and extended range of φ_c. (After Priebe, H.J., *Die Bautechnik*, 53(8), 1976.)

settlement contributions from layers of differing depths, properties, and stresses. This settlement s without soil improvement is then divided by the improvement factor β as taken from Figure 4.14 for the respective replacement ratio a_c and the column friction angle φ_c. The settlement with soil improvement s_i follows accordingly:

$$s_i = \frac{s}{\beta} \tag{4.16}$$

Stress concentration factors n after Priebe range between 5 and 11 for area replacement ratios a_c between 0.1 and 0.4 and for friction angles of the stone column material between 35° and 45°. These n values overestimate the effect of the stone columns to a certain extent when compared with field measurements in which n was found to be between 1.5 and 5, as already mentioned.

Priebe assumes that the stress state in the soil surrounding the stone column will be influenced by its installation in such a way that hydrostatic stresses develop. Consequently, he sets the earth pressure coefficient to $K_S = 1$. Other key assumptions initially made in this simplified calculation procedure are that the stone columns are based on a competent bearing stratum and that the column material is incompressible. In addition, both column and soil density are neglected, which has the consequence that the lateral support provided by the surrounding soil would not increase with depth, resulting in a column expansion that would be constant over its full length. Subsequent revisions of the method (Priebe, 1995, 2003) consider the positive influence of the compressibility of the column material, of the bulk weight of column and soil, and allow for an estimate of the additional settlement of *floating* stone columns that are not founded on a load-carrying stratum. One of these revisions (Priebe, 1988) also includes charts for computing the settlements of single and strip footings. Unfortunately, the settlement computations with these refinements are somewhat more cumbersome, as can be seen in Section 4.6.1, where a computational example demonstrates the use of the revised Priebe method. The method has also been frequently adapted to spreadsheet programs and ready-made software solutions. For a quick assessment of the expected settlements of a structure, the standard method is, however, considered adequate.

Goughnour and Bayuk (1979b) and Goughnour (1983) propose an incremental solution, based on the unit cell concept, for the settlement calculation of an infinite column grid. The unit cell is divided into slices of thickness Δh for which the vertical deformation is independently calculated, based on elastic and elastic–plastic analysis, with the larger of the two being the relevant deformation. Computation starts with the incremental slice at the column head and is carried on to the bottom of the column. Figure 4.15 shows a flowchart of the necessary computational steps. The method is, however, rather unsuitable for a conventional hand calculation because of the iterative calculation approach necessary for the solution of its set of equations. Computer programs were therefore developed instead.

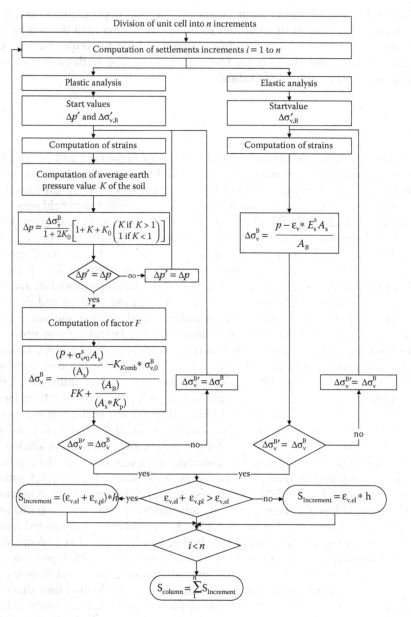

Figure 4.15 Computational steps of the Goughnour and Bayuk method.

Key assumptions of this method are that shear stresses are zero at the unit cell perimeter, that they are also zero at the surface and bottom of each incremental slice, and that no relative deformation occurs between column and soil. In Section 4.6.1, a sample calculation of this method in comparison with the Priebe method is presented.

From the relatively large number of methods proposed for the estimation of settlements of soft soil improved by stone columns, those have been proposed that are widely used today, delivering reasonable results within the normal accuracy bound of settlement calculations, provided that reasonable assumptions are made with regard to the shear strength of the column material and in particular to the area ratio of the column grid.

Numerical calculations adopting the finite element method are recommended for major projects and in the presence of particularly soft soils, say below $c_u = 20$ kPa, provided that the elastic–plastic behavior of the unit cell and the shear zones in the stone column can be modeled with sufficient accuracy. For large foundations, two-dimensional calculations may suffice, but other foundations may require the use of three-dimensional approaches.

4.3.3 Failure mechanism and bearing capacity calculations

The bearing capacity of a foundation on vibro replacement stone columns is governed by a number of possible failure mechanisms. For isolated columns these are

- Bulging of long, supported columns
- Shearing of short, supported columns
- Punching failure by sinking of short, floating columns
- Bulging in deeper layers

as demonstrated in Figure 4.16.

Bulging of the column is the main load-carrying mechanism of a stone column, which will fail when the surrounding soil cannot lend any further lateral support. In homogeneous soil conditions, bulging is concentrated within a region between the column head and a depth of about four times the column diameter. Short columns with lengths below four column diameters may fail by sinking into the soil. However, if these columns are placed on a reliable bearing stratum and the lateral support by the soil is sufficient, failure surfaces will develop through the column head and the adjacent soil close to grade. The described failure mechanisms are idealized. Failure may also occur at greater depths than four column diameters in the presence

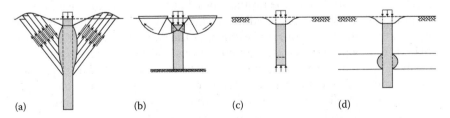

(a)   (b)   (c)   (d)

Figure 4.16 Failure mechanisms of isolated stone columns: (a) bulging, (b) shearing, (c) sinking, and (d) bulging in soft deeper layer.

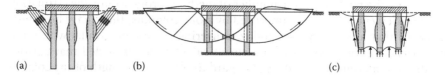

(a)   (b)   (c)

Figure 4.17 Failure mechanisms for column groups: (a) bulging, (b) shearing, and (c) bulging and sinking.

of particularly soft soil layers that are thick enough (say in excess of two column diameters) to allow bulging to develop as shown in Figure 4.16d.

Column groups show basically similar failure mechanisms, which are however less simple due to the complexity of the interaction between load application, soil and columns, and their geometrical parameters. Figure 4.17 displays failure mechanisms for column groups that can develop with rigid concrete foundations.

When looking at the infinite stone column pattern, the behavior of the circular unit cell under load has to be analyzed. In this symmetrical state of contained compression, no shearing will occur in the soil. Initially, soil and column will settle by elastic deformation and consolidation. However, the column is already under relatively low stresses, and will shear, dilate, and bulge, but it will be contained in plastic equilibrium, with the internal stresses corresponding with their plastic state. Typically, such infinite grids of stone columns are used for the foundation of storage tanks and embankments. To avoid uneconomical stone column patterns, design load on the unit cell needs to be high enough for the columns to pass through their maximum stiffness, rendering sufficient settlement reduction for the structure (Greenwood and Kirsch, 1984).

Similar conditions prevail for vibro stone columns below embankments where, besides reducing the overall settlement, they act predominantly to enhance slope stability. As will be shown later, the approaches that do not consider the stress concentration in the stone columns and that rely only on the higher shear resistance of the column material in relation to the native soil lead to overconservative factors of safety, and uneconomical designs. Design is generally based on conventional slip circle analyses and must take into account that, in order to fully mobilize the high column shear strength, substantial overburden pressure is required.

The easiest estimate of the ultimate bearing capacity of an isolated stone column makes use of a plane failure surface as defined by Bell (1915), and defines the equilibrium of a soil element at the column perimeter. The estimate is only valid for plane strain conditions and thus underestimates the capacity when applied to the axis-symmetric stone column. The maximum horizontal stress within the soil $\sigma'_{s,h}$ is governed by its undrained shear strength c_u and the vertical stresses q, which represents the loading on the soil next to the column when the soil itself is assumed to be weightless:

$$\sigma'_{s,h} = q + 2c_u \qquad (4.17)$$

The maximum vertical loading $\sigma'_{c,v,max}$ on the column can then be calculated using the passive earth pressure coefficient for the horizontal equilibrium as follows:

$$\sigma'_{c,v,max} = \sigma'_{s,h} \cdot \tan^2\left(\frac{\pi}{4} + \frac{\varphi_c}{2}\right) = (q + 2c_u) \cdot \tan^2\left(\frac{\pi}{4} + \frac{\varphi_c}{2}\right) \quad (4.18)$$

Based on the hypothesis of Greenwood (1970) that loaded single columns are in triaxially confined compression, all respective computational methods use the following relationship:

$$\sigma_{ult} = \sigma'_{c,v,max} = \tan^2\left(\frac{\pi}{4} + \frac{\varphi_c}{2}\right) \cdot (\sigma'_{s,h} + k \cdot c_u + u) \quad (4.19)$$

where:

$\sigma'_{c,v,max}$	is the maximum vertical column stress in kPa
φ_c	is the friction angle of column material
$\sigma'_{s,h}$	is the horizontal soil stress before column construction in kPa
k	is the factor of influence
c_u	is the undrained shear strength of soil in kPa
u	is the pore water pressure at column perimeter in kPa

The equation describes the failure state in a column according to Figure 4.16a, bulging, when the maximum horizontal stress in the column cannot be balanced anymore by the resisting horizontal soil stress. According to Brauns (1978), the factor of influence k can be computed by

$$k = 1 + \ln\left(\frac{E_{oed,s}}{3 \cdot c_u}\right) \quad (4.20)$$

with $E_{oed,s}$ being the oedometric, or constraint, modulus of the soil.

Brauns also proposes a relationship for the shear failure of a vibro stone column according to Figure 4.16b, which may occur near the column head as

$$\sigma'_{c,v,max} = \left(q + \frac{2 \cdot c_u}{\sin(2 \cdot \delta)}\right) \cdot \left(1 + \frac{\tan(\pi/4 + \varphi_c/2)}{\tan\delta}\right) \cdot \tan^2\left(\frac{\pi}{4} + \frac{\varphi_c}{2}\right) \quad (4.21)$$

where:

q is the surcharge at ground surface in kPa
δ is the angle of the assumed failure cone in the column

The minimum value of $\sigma'_{c,v,max} = \sigma_{ult}$ needs to be found by variation of the slip angle δ, with $\delta = 65°$ being a reasonable starting value.

Since no bearing capacity failure can occur with the infinite column grid, we look now at the failure of column groups under load as shown in Figure 4.17. The complexity of an analytical problem solution requires simplifying assumptions to be made with all methods available to date (Aboshi et al., 1979; Priebe, 1995), which are not always physically justifiable. It is therefore proposed to utilize the method after Barksdale and Bachus (1983), assuming planar failure surfaces within the column group.

The failure load q_{ult} is calculated as follows, utilizing the notations of Figure 4.18:

$$q_{ult} = \sigma_3 \cdot \tan^2 \delta + 2 \cdot c_{avg} \cdot \tan \delta \qquad (4.22)$$

where:

σ_3 is the average horizontal soil pressure in kPa
δ is the inclination of failure plane of the composite soil
c_{avg} is the average cohesion in kPa

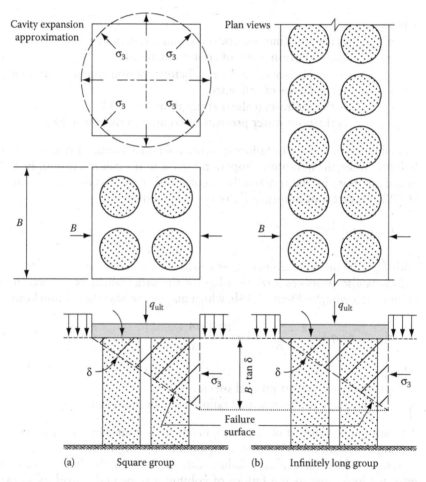

Figure 4.18 Bearing capacity of vibro stone column groups (a) below square footings and (b) long strip footings. (Based on Barksdale and Bachus, Design and Construction of Stone Columns. FHWA/RD-83/026, US Department of Transportation, Georgia Institute of Technology, Atlanta, GA, 1983.)

In this consideration, quick loading conditions are assumed with the undrained shear strength c_u prevailing in the soil ($\varphi_s = 0$) and with no cohesion for the stone column material. For the composite material of soil and stone column, an average friction angle φ_{avg} and a composite cohesion c_{avg} can be calculated using the Equations 4.26 and 4.28 below with n_c representing the stress concentration in the column. For simplicity, the stress concentration may be neglected with $n_c = 1$ and $\tan \varphi_{avg}$ calculated as the weighted average of the stone column and soil areas. However, this leads to conservative ultimate bearing capacities. With the improvement factor β developed with the Priebe method at the average depth of $\frac{1}{2} \cdot B \cdot \tan \delta$, the stress concentration factor n can be calculated using Equation 4.12 and introduced into Equation 4.27.

The lateral average earth pressure σ_3 is calculated by the classical earth pressure theory for the long strip foundation using Equation 4.23 and according to the cylindrical cavity expansion approximation after Vesic (1972) for the square foundation with Equation 4.24.

$$\sigma_3 = \frac{\gamma_s \cdot B \cdot \tan \delta}{2} + 2 \cdot c_u \tag{4.23}$$

$$\sigma_3 = c \cdot F_c' + q \cdot F_q' \tag{4.24}$$

where:

c	is the cohesion
q	is the mean stress $1/3 \cdot (\sigma_1 + \sigma_2 + \sigma_3)$ at failure depth
F_c', F_q'	are the cavity expansion factors according to Figure 4.19
I_r	is the rigidity index $E_s/(2(1 + \mu)(c + q \cdot \tan\varphi_s))$
E_s	is the Young's modulus of the soil
μ	is the Poisson's ratio in soil
φ_s	is the friction angle in soil
γ_s	is the unit weight of soil

The relevant slip angle δ can be calculated from Equation 4.25:

$$\delta = 45° + \frac{\varphi_{avg}}{2} \tag{4.25}$$

with

$$\tan \varphi_{avg} = \frac{n \cdot \tan \varphi_c}{(A/A_c) + (n-1)} = n_c \cdot a_c \cdot \tan \varphi_c \tag{4.26}$$

(remember $\varphi_s = 0$ in quick loading conditions) and

$$n_c = \frac{n}{(1 + (n-1) \cdot a_c)} \tag{4.27}$$

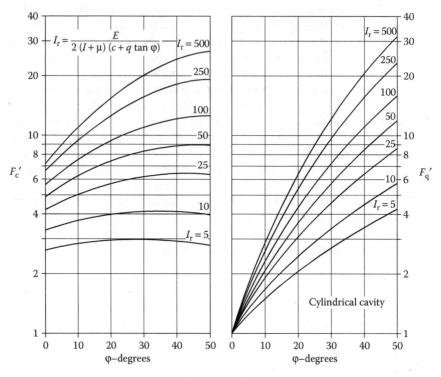

Figure 4.19 Cylindrical cavity expansion factor for soils with friction angle ϕ_s and cohesion c. (After Vesic, A.S., *JSMFD*. ASCE, 98, 1972.)

The average cohesion of the composite soil follows from

$$c_{avg} = (1 - a_c) \cdot c \tag{4.28}$$

A sample calculation of this method is shown in Section 4.6.3.

We have seen that a bearing capacity failure of an infinitely distributed, evenly loaded grid of stone columns cannot occur. However, at the edge of large column groups (below tanks and embankments), the general rotational or linear type shear failure resistance needs to be investigated. The design is generally based on conventional slip circle analysis and it is evident that, for a significant improvement of the safety factor, it is necessary to consider the stress concentration on the columns to fully mobilize their superior friction strength together with a substantial overburden pressure. Barksdale and Bachus (1983) have proposed, for this purpose, how to make use of the stress concentration n in hand calculations when adopting the unit cell concept.

Figure 4.20 shows such a unit cell at a depth where it intersects with the assumed failure line. The vertical effective stress $\sigma'_{v,c}$ resulting from the column weight and the applied stress σ can be expressed by

$$\sigma'_{v,c} = \gamma_c \cdot z + n_c \cdot \sigma \tag{4.29}$$

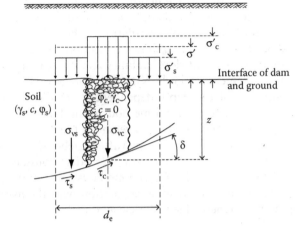

Figure 4.20 Unit cell concept for slope stability analysis. (After Barksdale, R.D. and Bachus, R.C., Design and construction of stone columns, FHWA/RD-83/026, US Department of Transportation, Georgia Institute of Technology, Atlanta, GA, 1983.)

with γ_c being the unit weight of the stone column material (use buoyant weight in submerged condition) and n_c representing the stress concentration factor in the stone column at depth z.

The shear strength of the stone column follows then as

$$\tau_c = (\sigma'_{v,c} \cdot \cos^2\delta) \tan\varphi_c \tag{4.30}$$

Vertical effective stress $\sigma'_{v,s}$ and shear stress τ_s in the soil surrounding the column are calculated from

$$\sigma'_{v,s} = \gamma_s \cdot z + n_s \cdot \sigma \tag{4.31}$$

and

$$\tau_s = c + (\sigma'_{v,s} \cdot \cos^2\delta)\tan\varphi_s \tag{4.32}$$

The weighted average unit weight γ_{avg} of the reinforced unit cell is used to calculate the acting moment.

$$\gamma_{avg} = a_c \cdot \gamma_c + a_s \cdot \gamma_s \tag{4.33}$$

With the known stress distribution factors n which have to be estimated or which can be calculated relatively easy by the Priebe method using the relationship between n and β according to Equation 4.12 and taking β from Figure 4.14, the slope stability analysis can be carried out with known standard methods. Calculating n, n_c, and n_s for different depths z from Equations 4.6, 4.8, and 4.9 using Priebe's graphs for the improvement factor β can be avoided by opting for an average depth z_{avg} taken as representative of the slip circle or plane considered, and establishing in this way the average stress concentration factors. An example calculation of the stability of an embankment founded on stone columns is shown in Section 4.6.2.

When comparing results of slope stability analyses carried out by three-dimensional finite element methods (Kirsch and Sondermann, 2003)—with a good reproduction of the depth-dependent stress distribution between column and soil with conventional methods using homogenization of the shear parameters by the weighted area average or the Barksdale and Bachus method with the β and n values developed after Priebe—it can be concluded that the weighted area method underestimates safety considerably, that the combined Barksdale and Bachus/Priebe methods overestimate safety, and that a safety factor obtained from a calculation with the average of both methods represents a reasonable result close to reality.

4.3.4 Drainage, reduction of liquefaction potential, and improvement of earthquake resistance

It is obvious that stone or gravel columns installed in fine-grained and cohesive soils represent elements of considerably greater permeability than the surrounding native soils. In water-bearing soils their beneficial effect of accelerating consolidation settlement and reducing the liquefaction potential of sandy soils (Seed and Booker, 1977) is generally appreciated.

As with sand drain construction, the wet installation of vibro replacement stone columns is superior to the displacement dry methods because it avoids the development of smear zones along the column perimeter. For standard, nonseismic applications, the choice of grading of the stone column material installed by the wet process in sand, silty sand, and sandy silt is unproblematic even when using uniformly graded coarse backfill, since the voids are generally filled with the remaining coarser particles of the native soil which were not washed out during column installation, in this way precluding material entry from the surrounding area.

When looking at the gradation of the column material for vibro replacement stone columns installed by the dry method—which is generally quite uniform as we have seen in Section 4.1—the question of filter stability is occasionally raised. It is suggested that the uniform grading of the column material would allow surrounding cohesive material to be transported with the seepage flow into their voids, thereby not only clogging them up but also leading to additional settlement. This effect has never actually been observed in practice, and dry vibro replacement stone columns generally show similar load settlement behavior to wet columns. When analyzing this problem, the Terzaghi filter rule cannot be used as it is only applicable for noncohesive soils. Stability of the boundary between the granular column material and the cohesive soil is governed instead by the relevant void diameter of the filter, the tensile strength of the cohesive soil, and the prevailing hydraulic gradient. It appears from a recent study of this question that these hydraulic gradients, measured in practice, are too small to cause material transport into the stone columns (Boley, 2007).

In water-bearing nonplastic silts that generally behave like granular soils, the choice of method (dry or wet) and column material deserves more careful consideration. Modern dry bottom feed systems allow the use of well-graded gravel and even sand as column material to avoid any transport of fines into the columns as a result of seepage flow. However, this choice would result in a distinct reduction of the friction angle of the column material.

In water-bearing fine-grained soils under load, that are improved by stone columns, pore water moves toward columns and drainage layers above or below. The curved flow paths can be divided into vertical and radial components, and the seepage flow rules adopted accordingly. Equation 4.34 describes the degree of consolidation U in the stone column-reinforced layer as the combined effect of the vertical consolidation U_v and the radial consolidation U_r.

$$U = 1 - (1 - U_v) \cdot (1 - U_r) \tag{4.34}$$

The settlement s_t at time t is in direct proportion with the degree of consolidation U and the final settlement s.

$$s_t = U \cdot s \tag{4.35}$$

The time factor for vertical flow T_v is given by Equation 4.36 as a function of the elapsed time t, the length of the drainage path h, and the coefficient of vertical consolidation c_v. The time factor T_r for radial flow is given by Equation 4.37. Figure 4.21 gives the notations used for vertical and radial flow conditions, in the latter case utilizing the unit cell concept. Figure 4.22 provides the degree of consolidation U_v and U_r as a function of the time factor T_v and T_r for vertical (v) and radial (r) flow conditions with stone column grid parameters $d_e/d = 3, 5,$ and 10.

$$T_v = \frac{c_v \cdot t}{h^2} \tag{4.36}$$

$$T_r = \frac{c_h \cdot t}{d_e^2} \tag{4.37}$$

$$\frac{c_h}{c_v} = \frac{k_h}{k_v} \tag{4.38}$$

where:

   k_v and k_h are the vertical and horizontal permeability of the soil, respectively

   c_v and c_h are the coefficient of vertical and horizontal consolidation, respectively

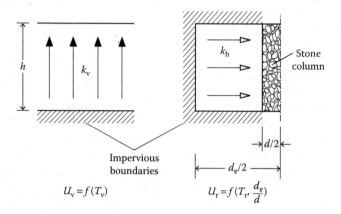

Figure 4.21 Vertical and radial drainage flow characteristics.

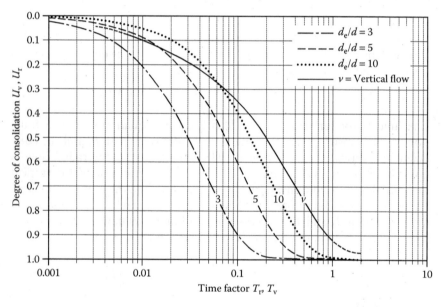

Figure 4.22 Time factors T_v and T_r for vertical and radial seepage flow. (Based on Terzaghi and Peck, *Die Bodenmechanik in der Baupraxis.* Springer-Verlag, Berlin, Germany, 1961 and Barron, *Transactions of ASCE,* 113(2346), 718–742, 1948.)

With known parameters for the above expressions, the consolidation time t can be calculated for different degrees of consolidation of the composite layer of soil and stone columns. In practice, it is helpful to remember that horizontal drainage is dominant in this scenario. Not only is the length of the flow path within an infinite stone column pattern in radial direction very often considerably shorter than in the vertical direction, but also horizontal permeability in clays is generally up to 15 times larger than in vertical direction. Consequently, it is sufficient to assess the consolidation time by neglecting the effect of vertical consolidation and to rely only on radial consolidation.

The relevance of this proposal can easily be verified by the following consideration. Let us assume that an infinite pattern of stone columns with diameter $d = 1$ m is placed in a triangular grid with distances of $a = 3$ m, yielding the unit cell diameter $d_e = 3.15$ m. Column length is 10 m, equal to the vertical drainage path length. Let us calculate the degree of combined consolidation at time t when the vertical consolidation U_v has reached just 10%. Assuming conservatively a ratio of 1 for horizontal to vertical permeability, and using Equations 4.36 and 4.37 at time t, we get an expression for the time factor of radial consolidation T_r as follows:

$$T_r = \left(\frac{h}{d_e}\right)^2 \cdot \left(\frac{c_h}{c_v}\right) \cdot T_v \qquad (4.39)$$

and with (Equation 4.38)

$$T_r = \left(\frac{h}{d_e}\right)^2 \cdot \left(\frac{k_h}{k_v}\right) \cdot T_v = \left(\frac{10}{3.15}\right)^2 \cdot T_v = 10.08 \cdot T_v \qquad (4.40)$$

From Figure 4.22, we obtain, for $U_v = 10\%$, the time factor $T_v = 0.01$ for vertical consolidation, which when used in Equations 4.39 and 4.40 leads to $T_r = 0.1$. From Figure 4.22 follows now with $d/d_e = 3.15 = {\sim}3$, the degree of radial consolidation $U_r = 90\%$. With Equation 4.34 the combined consolidation ratio is then $U = 91\%$. Accordingly, at the chosen time t, the contribution of vertical consolidation represents only 1% of the combined total consolidation.

$$U = 1 - (1 - 0.1) \cdot (1 - 0.9) = 0.91 \qquad (4.41)$$

This example was used to demonstrate that the vertical drainage flow can be neglected under normal conditions since radial consolidation generally governs the rate of combined consolidation of soil improved by stone columns. This also shows that this simplification is a safe assumption.

When adopting the dry bottom feed method for the stone column construction, a zone of reduced column permeability may develop at its perimeter as a result of mixing of disturbed soil with the column material. In order to account for this *smear effect*, as it is called, it is recommended to reduce the column diameter by 5% and to use this effective diameter in the relationships presented above. Regardless of the construction method, it is important to ensure the vertical flow of water in the stone columns, and this requires a permeable granular blanket to be placed over the column heads on ground surface that has good horizontal drainage characteristics.

When stone columns are installed in nonliquefiable cohesive soils, they can take considerable horizontal earthquake loads. These horizontal loads are generally transferred by friction from the load distribution layer placed below the foundation into the heads of the stone columns. When considering only the shear capacity of the stone column, that is, neglecting the higher shear resistance of stone column and tributary soil, a safety factor against earthquake-induced shear failure can be defined for the infinite stone column pattern with the notations from Figure 4.23 according to

$$SF = \frac{\tau_c \cdot A_c}{a_h \cdot \sigma_v \cdot A} = \frac{\sigma_c \cdot \tan\varphi_c \cdot A_c}{a_h \cdot \sigma_v \cdot A} = \frac{n_c \cdot a_c \cdot \tan\varphi_c}{a_h} \tag{4.42}$$

where:

τ_c is the shear resistance which can be mobilized in the stone column
a_h is the horizontal design earthquake acceleration factor
σ_v is the acting normal foundation stress
a_c is the area replacement factor
n_c is the stone column stress concentration factor

The allowable shear stress τ_c developed by the stone column can either be verified by a direct shear test executed on an actual column on-site (see Figure 2.15 and Section 4.4), or can be estimated from the relationship for n_c in which the area replacement factor a_c and the stress concentration factor were introduced:

$$\tau_c = \sigma_c \cdot \tan\varphi_c = n_c \cdot \sigma_v \cdot \tan\varphi_c \tag{4.43}$$

When introducing Equation 4.43 into 4.42, the safety factor can be rewritten as a function of the angle of internal friction φ_c of the stone column material and the horizontal design earthquake acceleration factor a_h. In cases where

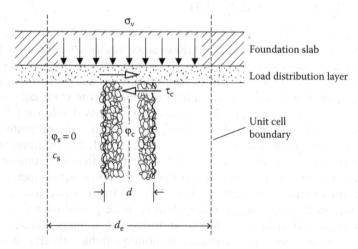

Figure 4.23 Shear forces acting on unit cell as a result of earthquake motions.

the vertical component a_v of the ground acceleration also needs to be considered, the resisting normal stresses σ_c in the stone column would have to be reduced by the vertical design earthquake acceleration a_v accordingly and the expression in Equation 4.42 would then be

$$SF = \frac{a_v \cdot n_c \cdot a_c \cdot \tan\varphi_c}{a_h} \tag{4.44}$$

The column stress concentration factor n_c is in the order of 2 close to foundation level. It can also be estimated with the assumptions of the Priebe method by first taking the improvement factor β from Figure 4.14 for a_c and φ_c, then calculating n and n_c with Equations 4.12 and 4.6. However, as previously indicated, the Priebe method seems to overestimate the effect of the stone column leading to n_c values of about 3 for the infinite grid with $a_c = 0.25$ and $\varphi_c = 45°$. Unfortunately, only few data from field measurements exist for n and n_c but they indicate n_c values between 1.5 and 2.0 only (Kirsch, 2004). It is therefore recommended to conduct field shear tests in critical cases to establish the allowable shear force that can be mobilized under actual conditions. In Section 4.6.4, a detailed discussion of the stress concentration factors can be found which will help to provide a better understanding of the variation of this parameter with depth t and relative column length λ.

Provided that sufficient compaction time is allowed, the stone column installation will also significantly increase the relative density of loose, clean sand layers, which could be imbedded in the soils to be improved, and which might liquefy. To ensure that sufficient density was achieved to prevent liquefaction, the methods described in Section 3.3.3.1 need to be applied with the earthquake-induced settlements to be estimated according to Section 3.3.3.2.

However, in silty sands that can only marginally be compacted by vibro compaction, the stone columns will act as gravel drains and will help to prevent the detrimental pore water pressure build-up occurring in cohesionless soils during an earthquake. Seed and Booker (1976) have developed design principles for stone or gravel columns acting as drains by which the necessary drain spacing and diameter can be evaluated to avoid liquefaction. Figure 4.24 gives useful design charts allowing determining the necessary relative stone column pattern spacing d/d_e as a function of the maximum allowable pore water pressure increase r_g, a ratio for the earthquake severity N_{eq}/N_l, and the time factor T_{ad} according to the following expressions:

$r_g = u_g/\sigma_v'$ is the greatest pore water pressure ratio allowed in the design
N_{eq} is the equivalent numbers of uniform stress cycles causing a stress ratio τ_{av}/σ_v' during time t_d (see also Section 3.3.3.1)
N_l is the number of stress cycles causing initial liquefaction in the laboratory with τ_{av}

$$T_{ad} = \frac{k_h}{\gamma_w} \cdot \frac{t_d}{m_{v3} \cdot (d/2)^2}$$

(4.45)

where:

k_h is the horizontal permeability of the soil
γ_w is the unit weight of water
t_d is the duration time of earthquake
d is the diameter of drain
m_{v3} is the coefficient of volume compressibility

The above design assumes an infinite permeability for the stone column material but it is also valid for stone column permeability in excess of $200 \cdot k_h$ of the soil. In addition to the soil permeability k_h, the computation requires the following parameters, which needs to be determined specifically for each project:

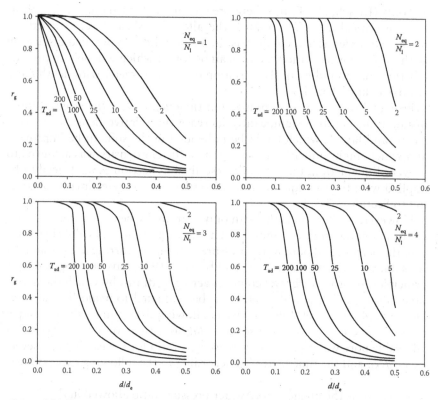

Figure 4.24 Design charts for the relative stone column spacing d/d_e as a function of the greatest design pore water pressure ratio r_g for different time factors T_{ad} and earthquake severities N_{eq}/N_l. (After Seed, H.B. and Booker, J.R., Stabilization of potentially liquefiable sand deposits using gravel drain systems, Report No. EERC 76-10, University of California. Berkeley, CA, 1976.)

- The number of equivalent stress cycles N_{eq} and the duration t_d of the design earthquake
- The liquefaction resistance of the soil expressed by the number of stress cycles N_1 causing liquefaction
- The allowable pore water pressure ratio r_g chosen for the design

With the calculated time factor T_{ad}, the relative stone column spacing can be determined from Figure 4.24 for various earthquake severities.

The selection of an adequate grading of the stone column material for seismic applications requires special considerations. It is important that the high permeability of the vibro stone column is maintained during the earthquake by a proper choice of the grading of the backfill material. Saito et al. (1987) have presented a particle size selection standard for the use of gravel in vertical drains as a countermeasure for sand liquefaction. The proposed filter criterion, which was developed for dynamic loading conditions with less restrictive lower limits for D_{15} of the filter material when compared with Terzaghi's filter rule is given by Equation 4.46.

$$20 \cdot d_{15} < D_{15} < 9 \cdot d_{85} \tag{4.46}$$

The notations D_{15} and d_{15} signify that 15% by weight of smaller grain diameters of the filter material and the natural soil, respectively, are passing the sieve. Vrettos and Savidis (2004) describe an interesting case history of the successful improvement of liquefiable silty sand by stone columns for the foundation of an immersed road tunnel in Greece. The marine stone column installation used gravel with a grain size distribution according to the Saito criterion for highly permeable, choke-free drains. In total over 130,000 lin.m of stone columns with a diameter of 600 mm and a nominal length of 15 m were executed from a barge to depths of up to 42 m applying the wet method. In Section 4.7.7, another example is described where the high permeability of stone columns helped building the foundations of a power plant in an earthquake-prone environment.

Ground improvement to mitigate the liquefaction potential of cohesionless soils by vibro replacement stone columns constructed using the dry bottom feed method has become a standard foundation solution in the United States. This method not only utilizes the compaction effect of the depth vibrator, but also combines it with the enhanced drainage capacity of the stone columns placed within the less permeable sand material. Stone backfill used in these instances generally consists of sufficiently hard, durable, clean, crushed rock, free of vegetable matter and other deleterious substances. The stone should have a sufficiently high durability index— over 40 measured by the California Test 229 as specified by the State of California, Department of Transportation. The gradation of the backfill suitable for the bottom feed gravel system and designed for an SW sand would be reflected by 100% passing the 50 mm sieve, 90%–100% passing the 25.4 mm sieve, and 10%–80% passing the 12.7 mm and up to 5% passing the 4.75 mm sieve. Should higher sand contents be required for economic or technical reasons, up to 30% passing the 4.75 mm sieve can be used. Typically, for these projects a maximum distance of the stone columns, (generally in the range of 3 m) together with their required diameter of about 800–900 mm would be specified. The sand between the columns would have to be compacted to certain equivalent clean sand CPT tip resistance values; in general, normalized and corrected for overburden pressure in accordance with the seismic risk evaluation (see Section 3.3.3 and the case histories in Sections 3.6.5 and 4.7.8).

4.3.5 Recommendations

The various steps in designing a vibro replacement foundation will generally commence with an assumed and preliminary arrangement of the stone columns, which will have to be verified by the assessment of bearing capacity and deformation under load. These assumptions primarily concern attributable load per column (generally between 200 and 500 kN), distance between column axis (generally between 2 and 5 column diameters), stone column diameter (ranging between 0.5 and 1.2 m), and column depth (ranging normally between 2 and 25 m). The choice will have to take into account the prevailing soil conditions, the available coarse backfill material and the chosen depth vibrator, and the method of installation. In water-bearing soils of low plasticity (sandy silts to clayey silts), depth vibrators with larger diameters will be used in general, and column construction will be by either the wet or the bottom feed dry method, always creating columns in excess of 0.8 m. Conversely, in stiffer cohesive soils, smaller diameter, high-frequency depth vibrators will ensure penetration to the required depth and will generally form columns with diameters below 0.8 m.

As we have seen when discussing the design principles, essential parameters controlling the performance of the stone column foundation are their diameter and the distance between each other, expressed by the area replacement factor a_c, together with the angle of internal friction φ_c of the column material. The gravel material which is normally used is generally coarse-grained alluvial deposits or crushed rock. It's stress-dependent friction angle usually decreases with increasing normal pressure, with $\varphi_{c,max}$ corresponding with $\sigma_{c,min}$ and vice versa. The friction angle of gravel is, as with other granular material, also directly proportional to its density. Installation methods generally guarantee that the column material is very well compacted by the vibratory motions of the depth vibrator with void ratios in general close to minimum values as field measurements have shown (Herle et al., 2007). Their shear parameter φ_c can be taken from a σ–τ diagram, which can be determined relatively easily by shear box tests in the laboratory. Maximum friction angles can be as high as 60° (generally measured at moderate pressures of ~50 kN/m²) with minimum values of about 50°

measured at normal pressures of over 200 kN/m². Herle et al. (2007) have also noted that conventionally used friction angle values of only 40° appear too conservative when designing vibro replacement stone columns.

Table 4.2 shows stress-dependent peak friction angles of dense gravel and provides useful guidelines for selecting the appropriate friction angle of the stone column material, where its physical properties can be considered and also the prevailing stress regime. We remember that a loaded stone column tends to bulge under load generally within its upper part. In this plastic state of equilibrium, it appears appropriate to use the residual friction angles in any settlement or slip failure calculation. It is therefore recommended to reduce the peak friction values in Table 4.2 by 5%–7% to obtain approximate residual values. Whenever backfill material with unknown shear characteristics has to be used it is strongly recommended to perform suitable laboratory tests for determining their friction angle for design purposes.

The material chosen for constructing the stone columns needs to be chemically inert to resist aggressive groundwater, and physically strong enough to endure the abrasive forces emanating from the vibrator during column installation. Their grain size distribution should enable the formation of a dense column and guarantee sufficiently high and permanent permeability for effective drainage. The European standards EN 14731 on deep vibratory methods and EN 1097–2 and EN 13450 on physical properties of crushed rock provide useful guidelines for specifying sufficiently durable stone backfill for use in stone column construction. Whenever recycled material is

Table 4.2 Stress-dependent friction angles of dense gravel to be used as stone column material

Type of gravel	$\varphi_{c, max}$ (°)	$\sigma_{c, min}$ (kN/m²)	$\varphi_{c, min}$ (°)	$\sigma_{c, max}$ (kN/m²)	Remarks
Crushed lime stone	63.1	50	53.8	200	DS
River gravel	58.8	50	51.9	200	DS
River gravel, subround	57.1	50	50.9	200	DS, $d_{60}/d_{10} = 2.6$
River gravel, subround	59.2	50	53.2	200	DS, $d_{60}/d_{10} = 2.1$
River gravel, crushed	60.4	50	55.2	200	DS
Basalt	71.8	8	45.6	240	TX, $D_{50} = 30$ mm
Basalt	70.0	8	51.1	120	TX, $D_{50} = 39$ mm
Basalt	64.2	27	45.6	695	TX
Sandstone	60.1	27	37.4	695	TX
Dolomite	64.0	15	43	500	TX, $\gamma = 1.7$ g/cm³
Dolomite	54.0	15	40	500	TX, $\gamma = 1.5$ g/cm³

Source: Data from Herle, I. et al., Einfluß von Druck und Lagerungsdichte auf den Reibungswinkel des Schotters in Rüttelstopfsäulen, in *Pfahl Symposium 2007*, Institut für Grundbau und Bodenmechanik. TU Braunschweig, Braunschweig, Germany, 84, 2007.

Note: TX, triaxial test; DS, direct shear test; and d_{60}/d_{10}, uniformity coefficient.

foreseen for this purpose, its abrasion resistance and its chemical characteristics require special consideration.

When adopting the displacement method of constructing stone columns, the horizontal stresses increase in the soil adjacent to the stone columns during their installation more significantly than with the wet replacement method. This stress increase is permanent and results also in an elevated modulus in those soils that are sufficiently stiff and do not tend to creep. Kirsch (2004) has measured this effect in a field test on two groups of 25 stone columns each in silty clay and sandy silt, respectively. Column diameter was $d = 0.8$ m and column depth 6–9 m. Figure 4.25a shows the effective horizontal stresses after column installation in relation to the initial stresses, expressed as the ratio of the earth pressure factors at rest K_0 after and before (index i) column installation, with a maximum of 160% (approx.) at a distance of between $4d$ and $5d$ from column axis. Figure 4.25b shows the modulus increase measured with the Menard pressure meter with a maximum of about 2.5 times the initial stiffness at distances of between 4 and 6 column diameters.

Figure 4.25 also shows the stress relief due to remolding caused by dynamic excitations in the vicinity of the columns, which will normally recover to original levels during reconsolidation of soils. The same paper also provides the results of a numerical analysis that simulates this installation effect on a foundation supported by 25 stone columns. Figure 4.26 summarizes the findings and compares the load settlement behavior of this footing, which is characterized by an area replacement factor (total stone column area/footing area) of $a_c = 0.28$, $\varphi_c = 45°$, and $l_c = 6$ m ($\lambda = 0.5$, *floating* situation), as derived from the numerical analysis with the results of standard analytical computations according to Priebe (2003) and Goughnour and Bayuk (1979b) (extended by Kirsch, 2004, to column groups).

It is interesting to see that the numerical method simulating the installation effect with groups of stone columns ties in well with the Priebe method (a) leading to improvement factors β that are about 50% higher than those calculated

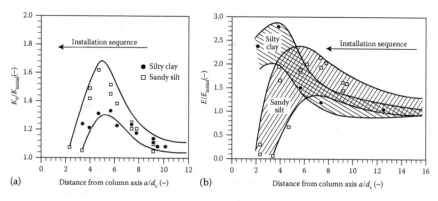

Figure 4.25 (a) Horizontal stress increase and (b) development of ground stiffness, during installation of stone columns.

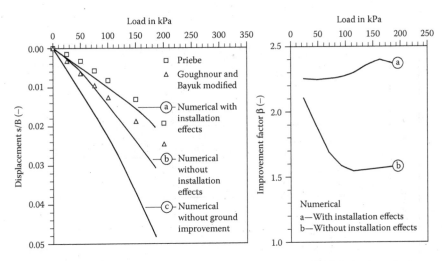

Figure 4.26 Analytical and numerical results of the load settlement behavior of a group of 25 stone columns ($a_c = 0.28$, $\varphi_c = 45°$, $\lambda = 0.5$, and settlement s normalized with foundation width B) and influence of column installation on the improvement factor.

without (b). This hidden reserve on the improvement factor as a result of the beneficial influence of the column installation in certain soils, as described above, indicates that the use of the standard analytical method after Priebe suffices for settlement estimation in such soils. Further studies to better define the characteristics of these soils in greater detail is nevertheless recommended.

Numerical simulations of the behavior of stone column groups are helpful to better understand their performance under load and to provide useful recommendations for their design. Figure 4.27 gives a flowchart for the various steps necessary for the design based on settlement.

The study has shown that only with sufficiently high replacement values (a_c between 0.2 and 0.5), with a friction angle of the column material φ_c of at least 45° and a relative stone column length λ of more than 0.5, improvement factors β between 1.5 and 2 can be achieved when dealing with column groups. Where the stone columns can be extended into a bearing stratum ($\lambda = 1$) the improvement factor increases to $\beta = 2.5$. Selection of a stone column material that, when properly compacted, allows friction angles φ_c of over 45° to be used will further increase the improvement factors to values in excess of 3.

As a result of the study, it can be concluded that stone columns should preferably always be extended into a competent stratum, and only in exceptional cases, when the thickness of the layer needing improvement is large in comparison with the width of the foundation, floating columns should be executed. However, in these cases, the stone columns should always be longer than 1.5 times the smallest width of the footing. If β needs to be increased further, it is more effective to achieve this by increasing the column length

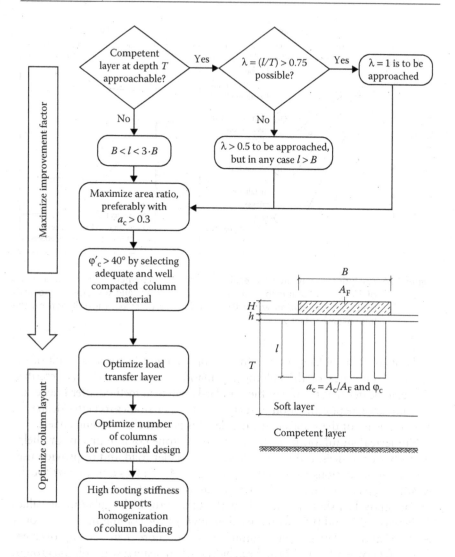

Figure 4.27 Flowchart for the design of a group of stone columns.

beyond $\lambda = 0.75$. If, however, the thickness of the soft soil is very large in comparison with the dimensions of the footing, stone column lengths in excess of three times the footing width no longer contribute to the settlement reduction. Any chosen design must finally be checked with regard to its bearing capacity according to Section 4.3.3.

Generally, a load distribution layer consisting of gravel or crushed rock with a thickness of at least 300 mm is placed over the vibro stone column heads. This layer also serves to safeguard the drainage capability of the stone columns in water-bearing soils when loaded, but requires effective access to a working recipient. Thorough compaction of this layer is recommended before concreting work commences. Measurements and the above parametric study reveal that this layer helps to concentrate foundation loads into the stone columns. It helps to equalize the bearing stresses acting on the footing. However, the load distribution layer itself does not increase the improvement factor of the stone column-reinforced strata any further.

Footing and slab stiffness help to evenly distribute foundation loads on the stone columns leading finally also to evenly distributed settlements. stone columns should always be arranged directly below footings and foundation slabs. Only when other criteria dictate—bearing capacity, slip failure, or earthquake mitigation—do they also need to be placed outside.

For the prevention of liquefaction, it is also recommended to extend the stone columns outside the structure over a width which is equal to at least two-thirds of the soil improvement treatment depth.

4.4 QUALITY CONTROL AND TESTING

In addition to the quality control measures that were discussed in Section 3.4 for the vibro compaction method, concerning assurance of the geometrical layout of the vibro stone column pattern and their treatment depth, verification of the column diameter d is of prime importance. This geometrical relation is represented by the area replacement factor a_c, which significantly influences design and performance of the vibro stone column foundation.

Modern data acquisition systems for vibro replacement collect, store, and display all relevant data of a stone column including column identification number and, as a function of real time, depth, AC current or hydraulic pressure of the vibrator motor, stone volume (measured directly or deduced from measured stone weight), air pressure, and installation duration. All data are generally collected numerically and can be stored or used for graphical displays to support construction and supervision procedures or to serve for quantity surveying purposes (Bauldry et al., 2008). Interpretation of the graphical printouts obtained in this way is very important but requires experience. Deficiencies of the column installation, particularly necking of the column diameter, can however only be detected when real-time recordings of the parameters are made available.

Where the stone backfill quantity cannot be measured directly versus depth and recorded by the monitoring device, a realistic estimate of the total volume of stone needs to be carried out. It should be correlated with the total length of stone columns executed during reasonable time intervals and calibrated with the delivered stone quantity. Material wastage and excessive heave developing at the working platform during column installation have to be considered.

Where heterogeneous site conditions prevail and compactable sand layers alternate with cohesive soil layers, improvement can be achieved by a combination of the densification effect of vibro compaction with the rein-forcement effect of stone columns. In these instances, the degree of compaction can be measured in the sand layers by the methods described in Section 3.4. Improvement of the cohesive soils is then achieved in the first place by a sufficiently high area replacement value a_c and the friction angle φ_c of the stone column material.

The stiffness of the improved soil is controlled primarily by the shear strength of the stone column material. It is therefore recommended to verify, for each major site, the friction angle φ_c of the stone backfill chosen in the design with suitable laboratory tests, if insufficient experience with the material exists. In exceptional cases, and if the size of the project merits the expenditure, the shearing resistance of the stone column alone or of stone column and tributary soil (i.e., the unit cell) can also be measured directly on-site by full-scale shear tests (see Figure 2.15). Barksdale and Bachus (1983) recommended these shear tests to be performed as double ring tests as described in the Jourdan Road Terminal Test Embankment report (Parsons-Brinkerhoff et al., 1980).

Suitability of the stone backfill material in terms of hardness and abrasion resistance may be controlled by relevant testing methods such as the Los Angeles Test following ASTM C131 designations or similar regulations such as the European standards EN 1097–2 and EN 13450.

When verification of the performance of single or strip foundations on column groups is needed, full-scale loading tests are recommended with actual footing dimensions because of the complex interaction between stone columns, footing size, and stiffness. Alternatively, it may be possible to achieve sufficient confidence in some cases by performing tests with a footing on at least three columns.

For verification of the settlement performance of large column groups or infinite column grids, zone tests are necessary for modeling this loading scenario realistically. Load tests on single columns with the load only being applied directly on the column itself do not reflect actual stress conditions in these situations.

In cases where the soft soil to be improved is close to foundation level, say up to five times the unit cell diameter d_e, load tests on single columns are, however, indicative for the performance of an infinite column grid with the test footing size being equal to the unit cell area $A = 0.25 \cdot \pi \cdot d_e^2$. From the load-settlement curve of this unit cell loading test, an equivalent Young's modulus E^* can be defined as the ratio of the uniform load q multiplied by an equivalent column length l^* and divided by the measured settlement s_m. The uniform load q on the infinite grid causes the settlement of the infinite grid s_g with the actual column length l and the equivalent modulus E^* as follows:

$$\frac{q \cdot l^*}{s_m} = E^* = \frac{q \cdot l}{s_g} \tag{4.47}$$

In a parametric finite element analysis it was shown that the equivalent column length l^*, which is just a calculating unit, is essentially independent of the actual column length of the load test and of the stiffness ratio of column material and surrounding soil. It depends predominantly on the grid geometry as represented by the unit cell diameter d_e and can be taken for different unit cell diameters from Figure 4.28a.

These unit cell load tests on single columns to represent the load settlement performance of the infinite grid were developed from parametric studies and are, however, not yet verified in practice. In this context it is important, as with other load tests, to avoid any disturbing influence of the load application system on the settlement during their execution. It is therefore recommended to perform the load test with a setup such as that shown in Figure 4.28b.

Reports on the execution of load tests on groups of vibro stone columns are numerous, of which only a limited number of more recent publications are included in the references. These follow general ASTM or similar procedures for quick load tests on spread footings to indicate bearing capacity of the foundation applying test loads of at least 150% design load. For settlement control purposes, long-term tests are required whereby it is necessary to allow a sufficient time span to elapse (generally 0.01 mm/min) before the

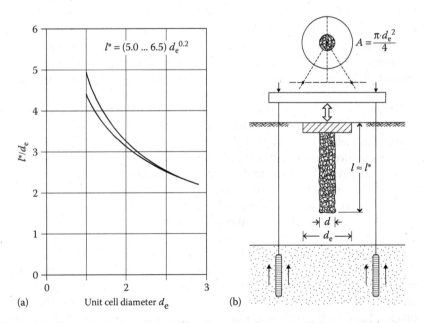

(a)   Unit cell diameter d_e (b)

Figure 4.28 Equivalent column length l^* as a function of the unit cell diameter (a) for determination of infinite column grid settlements from single column load tests (b). (After Kirsch, F. and Borchert, K.-M., Probebelastungen zum Nachweis der Baugrundverbesserungswirkung, in 21. *Christian Veder Kolloquium: Neue Entwicklungen der Bodenverbesserung*, TU Graz, Graz, Austria, 2006.)

next load increment is applied. Load test results need to be theoretically analyzed to predict as realistically as possible the actual performance of the stone column-reinforced ground under load. Unfortunately, field tests are expensive and very often for a true and realistic assessment of the effect of ground improvement certain elements—for instance the settlement performance of the untreated soil—are missing. It is therefore strongly recommended to carefully design all field measurements to reflect not only the relevant loading conditions as closely as possible but also the prevailing geotechnical conditions on-site.

Occasionally, static cone or dynamic penetration tests are proposed to measure the integrity (density and continuity) of the stone columns installed, although test results obtained in this way are often questionable if not misleading and generally difficult to interpret. Very often, the steel rod of the testing equipment deviates from the vertical and leaves the relatively slender stone column without noticeable indication, thereby giving room for incorrect interpretations. Scrutiny in checking the printouts of the data collection system is the better, and most commonly used way to detect deviations from the specified geometry of the constructed stone columns.

4.5 SUITABLE SOILS AND METHOD LIMITATIONS

In Figure 3.11, the application limits of the vibro compaction method are basically restricted to areas A and B. Soils with grain size distribution falling into areas C and D cannot be compacted anymore by vibratory motions, but their characteristics may be improved by the vibro replacement stone column method, provided that their stiffness does not prevent the depth vibrator penetrating further.

Degen (1997b) has provided useful comments for the applicability of the vibro stone column method in these soils using the USCS classification system (Table 4.3), which particularly take into account the plastic behavior of these soils. To securely construct a stone column, the soils must have a minimum strength, expressed by their undrained cohesion, which should not fall below $c_u = 5$ kPa, since otherwise it cannot provide sufficient containing pressure for the column. If these soils extend to ground level, their softness generally requires a competent gravel blanket of about 0.5 m thickness to be placed as a working platform over the whole site. Into stiffer soils at lower water content and cohesions of above $c_u = 50$ kPa, even slender, powerful, high-frequency depth vibrators may not be able to penetrate. Although their characteristics may not need improvement, softer soils may necessitate the stone columns to reach deep. In these cases, pre-drilling of the stiffer soils with suitable methods is advisable. Pre-drilling is often also necessary if a hard surface crust has developed by drying out or through site traffic, or when stone and weathered rock layers or other obstructions pose penetration problems.

In stiff soils with low water content, closely spaced vibro stone columns at distances below $3d$ can cause substantial heave at ground surface during

Table 4.3 Suitability assessment of fine-grained cohesive soils for vibro replacement

Soil type	USCS	Comment on suitability for vibro replacement stone columns
Silty sands	SM	Stone column necessary and suitable for silt content >10%
Clayey sands	stone column	Stone column with marginal overall compaction effect, very fast draining after treatment
Inorganic clays (low plasticity)	CL	$c_u \geq 5$ kPa recommended for upper 3 m, potential difficulties for vibrators to penetrate with $c_u > 50$ kPa (very stiff conditions)
Inorganic clays (high plasticity)	CH	As for CL, but not suitable when w_n too close to w_L
Silts and clays with $w_L < 50$	ML	Pre-boring necessary when dry
Inorganic and organic silts and clays with $w_L < 50$ and high plasticity	MH CH OH	Collapsing soils not suitable Soils generally not or only Marginally treatable
Peat, organic matter predominant	PT	In excess of 1 m thickness not recommended

Source: Modified from Degen, W., Vibroflotation Ground Improvement (unpublished), 1997b.

their installation. It is important in these circumstances to avoid neighboring columns causing mutual damage during construction. Should excessive heave be unavoidable, partial pre-boring can assist. After removal of the upper 0.5 m of surface soil following stone column installation, the site surface and stone column heads need to be recompacted before the load distribution layer is installed on top.

The practical depth range of vibro stone columns is 2–25 m. Exceptional cases may necessitate the installation of very deep stone columns, for example, if stability or liquefaction requirements dictate. For these cases, crane-hung bottom feed systems have been developed with depth capability, both on and offshore, in excess of 50 m. These not only need careful planning and thorough site investigations but also execution should only be by very experienced contractors.

The vibro stone column method is often used to treat heterogeneous strata, sometimes with successive layers of very different characteristics or with imbedded lenses of soft, organic soil of inconsistent extent and thickness. Site investigation will not always detect these ground conditions. Although the method is no substitute for a site investigation, the regular and well-documented installation method is likely to detect unexpected ground irregularities. The Engineer and contractor will then have to define ways and means of coping with the changed soil conditions, either by additional stone columns at closer spacing or by increasing the stone column diameter. In these circumstances stone columns may be able to bridge the soft layers if their thickness is not greater than about two column diameters. Any thicker layers will almost certainly have been detected by the site investigation and considered in the stone column design in the first place. The stone backfill may be replaced by concrete over the thickness of the soft organic layer while below and above the normal method of stone column construction is applied (see Chapter 5).

4.6 COMPUTATIONAL EXAMPLES

4.6.1 Analysis of settlement reduction

The following analytical example is based on geometrical and material properties, which were slightly idealized from an existing ground improvement project for the foundation of a road embankment. The soil is characterized by very soft alluvial deposits. These soils make vibro stone columns for infrastructural measures a very common technology, in order to reduce settlements. Figure 4.29 and Tables 4.4 and 4.5 show the cross section through the embankment and its foundation together with the necessary design parameters.

The settlement analysis as an analysis in the serviceability limit state according to Eurocode 7 is done first, using characteristic values of the soil parameters and applying the Priebe method (Section 4.3.2) for the center of the embankment, assuming unit cell conditions. The analysis requires the division of the soil mass into a sufficiently large number of horizontal layers. Here equidistant slices of 1 m thickness are chosen. The method of Priebe (1976) is based on a classical settlement analysis using stress distribution and the computation of a settlement improvement factor by which the calculated settlements are reduced.

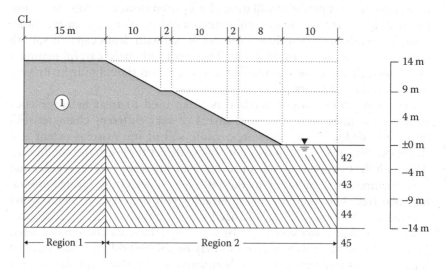

Figure 4.29 Cross section through embankment.

Table 4.4 Soil properties chosen in the design

Soil layer	γ/γ' (kN/m³)	E_{oed} (MN/m²)	φ' (°)	c' (kN/m²)	$\mu(-)$
1 embankment fill	21.8/12	—	35	0	0.33
42 firm silt	16/6	4	0	40	0.33
43 very soft silt	14/4	0.6	0	6	0.33
44 soft silt	14/4	0.8	0	8	0.33
45 hard silt	18/8	20	0	200	0.33

Table 4.5 Stone column parameters chosen in the design

Stone columns	Diameter (m)	Square grid spacing (m)	Area replacement $a_c(-)$	$E_{c,oed}$ (MN/m²)	γ/γ' (kN/m³)	φ_c' (°)
Region 1	1.1	2.1	0.22	120	19/12	42
Region 2	1.1	1.7	0.33	120	19/12	42

The basic improvement factor β_0, after Priebe, is calculated from

$$\beta_0 = 1 + \frac{A_c}{A} \cdot \left[\frac{1/2 + f(\mu_s, A_c/A)}{K_{ac} \cdot f(\mu_s, A_c/A)} - 1 \right] \tag{4.48}$$

where:

$$K_{ac} = \tan^2(45° - \varphi_c/2) \tag{4.49}$$

$$f(\mu_s, A_c/A) = \frac{1 - \mu_s^2}{1 - \mu_s - 2\mu_s^2} \cdot \frac{(1 - 2\mu_s) \cdot (1 - A_c/A)}{1 - 2\mu_s + A_c/A} \tag{4.50}$$

The basic improvement factor is usually reduced to account for the fact that a complete exchange of the soil by the column material should lead to a maximum improvement factor β_{max} represented by the ratio of the modulus of column material and the modulus of the soil.

$$\beta_{max} = E_c/E_s \tag{4.51}$$

Hence, the corrected improvement factor β_1 is calculated according to Priebe (1995) from

$$\beta_1 = 1 + \frac{\overline{A_c}}{A} \cdot \left[\frac{1/2 + f(\mu_s, \overline{A_c/A})}{K_{ac} \cdot f(\mu_s, \overline{A_c/A})} - 1 \right] \tag{4.52}$$

where

$$f(\mu_s, \overline{A_c/A}) = \frac{1 - \mu_s^2}{1 - \mu_s - 2\mu_s^2} \cdot \frac{(1 - 2\mu_s) \cdot (1 - \overline{A_c/A})}{1 - 2\mu_s + \overline{A_c/A}} \tag{4.53}$$

$$\frac{\overline{A_C}}{A} = \frac{1}{A/A_c + \Delta(A/A_c)} \tag{4.54}$$

$$\Delta(A/A_c) = \frac{1}{(A_c/A)_1} - 1 \tag{4.55}$$

with $\mu_s = 0.33$

$$\left(\frac{A_c}{A}\right)_1 = -\frac{4 \cdot K_{ac} \cdot (\beta_{max} - 2) + 5}{2 \cdot (4 \cdot K_{ac} - 1)}$$

$$\pm \frac{1}{2} \cdot \sqrt{\left[\frac{4 \cdot K_{ac} \cdot (\beta_{max} - 2) + 5}{4 \cdot K_{ac} - 1}\right]^2 + \frac{16 \cdot K_{ac} \cdot (\beta_{max} - 1)}{4 \cdot K_{ac} - 1}} \tag{4.56}$$

On the other hand, the improvement factor β_1 so calculated, which is mainly based on the assumption of a lateral support of the column by the surrounding soil, does not take into account the increasing support of the soil with depth t. Therefore, the improvement factor is increased by introducing the depth factor f_t to

$$\beta_2 = f_t \cdot \beta_1 \tag{4.57}$$

where

$$f_t = \frac{1}{1 + [(K_{0c} - 1)/K_{0c}] \cdot \Sigma(\gamma_s \cdot \Delta t)/p_c} \tag{4.58}$$

$$p_c = \frac{p}{A_c/A + [(1 - A_c/A)/(p_c/p_s)]} \tag{4.59}$$

$$\frac{p_c}{p_s} = \frac{1/2 + f(\mu_s, \overline{A_c / A})}{K_{ac} \cdot f(\mu_s, \overline{A_c / A})} \tag{4.60}$$

$$K_{0c} = 1 - \sin\varphi_c \tag{4.61}$$

The factor f_t is limited by

$$f_t \leq \frac{E_c/E_s}{p_c/p_s} \tag{4.62}$$

and the improvement factor β must not rise above β_{lim} given by

$$\beta_2 \leq \beta_{lim} = 1 + \frac{A_c}{A} \cdot \left(\frac{E_c}{E_s} - 1\right) \tag{4.63}$$

The above equations are best programmed, for example, in a spreadsheet, together with the equations for classical settlement calculation under load.

The results of the Priebe analysis are presented in Table 4.6. Ground pressure in the middle of the embankment is calculated by using

$$p = 14 \text{ m} \cdot 21.8 \text{ kN/m}^3 = 305 \text{ kN/m}^2 \tag{4.64}$$

Table 4.6 Settlement analysis with the Priebe method

Depth (m)	Settlement without improvement (mm)	Improvement factor β_2 (–)	Settlement with stone columns (mm)
1	76	2.38	32
2	76	2.38	32
3	76	2.38	32
4	76	2.45	31
5	508	2.50	203
6	508	2.50	203
7	508	2.50	203
8	508	2.50	203
9	508	2.50	203
10	381	2.49	153
11	381	2.49	153
12	381	2.49	153
13	381	2.49	153
14	381	2.49	153
15	15	1	15
16	15	1	15
Total	4783	2.47	1936

A second analysis shows the application of the method after Goughnour and Bayuk (1979b). It follows the flowchart as shown in Figure 4.15. As mentioned there, the algorithm consists of an iterative approach, which is best solved by using a computer routine.

The analytical procedure of Goughnour and Bayuk requires the definition of the initial void ratio e_0 and the compression index C_C. In order to ensure comparability with the above analysis, e_0 and C_C were chosen to fit the constrained modulus E_{oed} as shown in Table 4.4 in terms of settlements of the unimproved ground. Therefore the compression index was calculated from an assumed initial porosity $e_0 = 1.2$ by

$$C_C = (1 + e_0) \ln (10) \frac{\sigma_{v,0}}{E} \tag{4.65}$$

with the Young's modulus E and the initial vertical stress $\sigma_{v,0}$. All other parameters and the geometry were kept identical.

The Goughnour and Bayuk method calculates total settlements of the improved ground to be approximately 2.70 m (see Table 4.7) which is about 140% of the settlements calculated with the Priebe method. With a fully three-dimensional finite element analysis of the situation, a maximum

Table 4.7 Settlement analysis with the Goughnour and Bayuk method

Depth (m)	Compression index C_C	Initial void ratio e_0	Settlement without improvement (mm)	Settlement with stone columns (mm)
1	0.03	1.2	24	16
2	0.05	1.2	36	23
3	0.08	1.2	47	29
4	0.10	1.2	56	34
5	0.80	1.2	402	241
6	0.91	1.2	436	257
7	1.03	1.2	467	272
8	1.14	1.2	497	285
9	1.26	1.2	524	298
10	1.03	1.2	413	232
11	1.11	1.2	431	239
12	1.20	1.2	449	246
13	1.29	1.2	466	252
14	1.37	1.2	482	258
15	0.06	1.2	21	21
16	0.07	1.2	22	22
Total			4771	2726

settlement of 2.40 m was calculated in the center of the embankment (Kirsch and Sondermann, 2003).

4.6.2 Analysis of slope stability

The example from Section 4.6.1 will now be analyzed with respect to its stability. To this end, it is necessary to calculate the shear strength of the improved ground. In a first analysis the stability in the ultimate limit state (ULS) is calculated using characteristic values of both loads and resistances computing an overall factor of safety. Since the load is concentrated in the stone column material, it would be too conservative to compute average shear strength values only on the basis of the area ratio a_c, since allowable shear stresses in the column are governed by the stress-dependent inner friction. Therefore, Priebe (1995) suggests calculating the strength of the improved ground on the basis of ratio m of the load carried by the column to the total load acting on the unit cell.

$$m = \frac{\sigma_c \cdot A_c}{\sigma \cdot A}$$

(4.66)

With Equations 4.2 and 4.5, the above equation can be rewritten as

$$m = \frac{n \cdot a_c}{(1 + (n-1) \cdot a_c)}$$ (4.67)

Introducing Equation 4.12, we get

$$m = \frac{n \cdot a_c}{\beta} = \frac{\beta - 1 + a_c}{\beta}$$ (4.68)

Equation 4.68 denotes the maximum load ratio attributable to the stone columns. Conservatively, m can be reduced to m' by simply neglecting the area ratio a_c. The load ratio m_1' may then be calculated from the settlement improvement factor β_1 according to Priebe (Equation 4.52), which includes the correction for column compressibility by the following equation

$$m_1' = \frac{\beta_1 - 1}{\beta_1}$$ (4.69)

On the other hand, it might be unsafe to calculate the average shear strength based only on the load ratio of the stone columns, because doing so would apply the stress concentration also to the unit weight of the in-situ soil, which may not be conservative, especially with deep slip circles. In Kirsch and Sondermann (2003), it is therefore recommended to calculate the shear strength of the improved soil as the mean of the strength values on the basis of either the area ratio or the load ratio.

Priebe (2003) recommends modeling each individual stone column in a two-dimensional slip circle analysis with vertical slices, and factorizing the load on top of the columns by the stress concentration factor and simultaneously reducing the load on the soil in between. This can, for example, be achieved by adjusting the unit weight of the embankment. With large systems, this approach can be quite laborious. Therefore Priebe alternatively proposes calculating average shear strength values on the basis of the load ratio but reducing them with depth by applying a reduction factor being the quotient of the load on top of the improved ground Q_{top} and the total load Q_{total} at the respective depth. The maximum reduction of the load ratio is reached when the columns are unloaded. In this case, the strength of the improved ground is the average of the strength of the soil and the columns, weighted by the area ratio according to Priebe adopted for the compressibility of the column material:

$$m_1'' = \overline{A_c/A} + (m_1' - \overline{A_c/A}) \cdot \frac{Q_{top}}{Q_{total}}$$ (4.70)

With the adjusted load ratio m'_1, the average strength of the improved ground can be calculated by the following equations:

$$\varphi_{avg} = \arctan\left(m''_1 \cdot \tan\varphi_c + (1 - m''_1) \cdot \tan\varphi_s\right) \tag{4.71}$$

$$c_{avg} = (1 - m''_1) \cdot c_s \tag{4.72}$$

Calculating the average cohesion c_{avg} also on the basis of the reduced load ratio m'_1 is on the safe side, since physically one could apply the area ratio, which would lead to higher cohesion values.

In order to calculate the corresponding shear strength values, the system is divided into five vertical sections, A to E, in which the ground is improved by stone columns of different area ratios $a_c = 0.22$ in section A and $a_c = 0.33$ in sections B to E. Additionally, the four existing layers (Figure 4.29) are subdivided into sublayers of approximately 2 m thickness. The five sections A to E in Figure 4.30 are characterized by different loading conditions due to the different embankment heights which are averaged within the section for simplicity. Table 4.8 gives an overview of the resulting shear strength values. The soil properties of the unimproved ground (layers 42–45) and the embankment material (layer 1) are given in Table 4.4.

Figure 4.30 shows the embankment geometry with the different soil layers as chosen in the calculations, and the decisive slip circle, in this case with a minimum factor of safety of 2.08, together with the contours of equal safety.

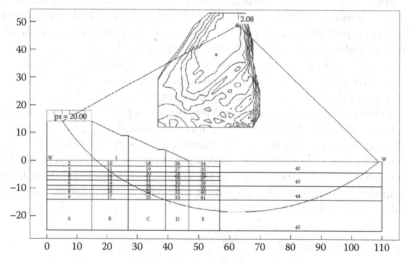

Figure 4.30 Cross section with soil layers as used in the computation and decisive slip circle.

Table 4.8 Depth-dependent shear strength of the improved ground in sections A to E

Depth (m)	Layer	m_1'	Q_{top}/Q_{total}	m_1''	φ_{avg} (°)	c_{avg} (kPa)	Layer	m_1'	Q_{top}/Q_{total}	m_1''	φ_{avg} (°)	c_{avg} (kPa)
			A						B			
−2.0	2	0.58	0.96	0.57	27.00	17.36	10	0.70	0.95	0.68	31.56	12.71
−4.0	3	0.58	0.93	0.55	26.47	17.88	11	0.70	0.91	0.67	30.95	13.36
−5.5	4	0.59	0.91	0.56	26.63	2.66	12	0.72	0.89	0.68	31.41	1.93
−7.0	5	0.59	0.89	0.55	26.39	2.69	13	0.72	0.87	0.67	31.14	1.97
−9.0	6	0.59	0.87	0.54	26.07	2.74	14	0.72	0.85	0.66	30.79	2.03
−10.5	7	0.59	0.86	0.54	25.84	3.70	15	0.72	0.83	0.66	30.54	2.76
−12.0	8	0.59	0.84	0.53	25.62	3.74	16	0.72	0.82	0.65	30.29	2.81
−14.0	9	0.59	0.83	0.53	25.34	3.79	17	0.72	0.80	0.64	29.98	2.87
			C						D			
−2.0	18	0.70	0.92	0.67	31.08	13.22	26	0.70	0.78	0.62	29.01	15.37
−4.0	19	0.70	0.86	0.64	30.09	14.26	27	0.70	0.64	0.56	26.82	17.54
−5.5	20	0.72	0.83	0.65	30.41	2.09	28	0.72	0.59	0.56	26.80	2.63
−7.0	21	0.72	0.80	0.64	29.99	2.15	29	0.72	0.55	0.54	26.08	2.74
−9.0	22	0.72	0.76	0.63	29.47	2.23	30	0.72	0.50	0.52	25.26	2.86
−10.5	23	0.72	0.74	0.62	29.10	3.05	31	0.72	0.47	0.51	24.74	3.91
−12.0	24	0.72	0.72	0.61	28.76	3.12	32	0.72	0.44	0.50	24.27	3.99
−14.0	25	0.72	0.69	0.60	28.33	3.21	33	0.72	0.41	0.49	23.72	4.10
			E									
−2.0	34	0.70	0.00	0.31	15.60	27.60						
−4.0	35	0.70	0.00	0.31	15.60	27.60						
−5.5	36	0.72	0.00	0.33	16.55	4.02						
−7.0	37	0.72	0.00	0.33	16.55	4.02						
−9.0	38	0.72	0.00	0.33	16.55	4.02						
−10.5	39	0.72	0.00	0.33	16.55	5.36						
−12.0	40	0.72	0.00	0.33	16.55	5.36						
−14.0	41	0.72	0.00	0.33	16.55	5.36						

The analysis is performed using a standard stability analysis program and applying a division into 50 slices. The groundwater table was set equal to the ground surface. An additional loading of 20 kN/m² was applied on top of the embankment.

According to Eurocode 7, partial safety factors on loads and resistances are necessary to be applied and it needs to be proven that the design value of the loads or actions E_d is smaller than the design value of the resistances R_d, or:

$$\frac{E_d}{R_d} \leq 1 \tag{4.73}$$

In order to compute the ULS of the slope stability, different design approaches (DAs) are allowed. In Britain, DA 1 is used, where two combinations of sets of partial safety factors need to be investigated. Alternatively, in Germany, DA 3 is utilized in which the actions—here the life load (e.g., from traffic) at the top of the embankment—are used as design (factored) values and the resistance in the resulting shear plane is calculated from designing (factored) shear parameters of the soil. In the above example all characteristic shear parameters φ_{avg} and c_{avg}, respectively, of Table 4.8 were factored according to the following equation:

$$\varphi'_{avg,d} = \arctan\left(\tan \varphi'_{avg,k}/\gamma_\varphi\right); \gamma_\varphi = 1.25 \tag{4.74}$$

$$c'_{avg,d} = c'_{avg,k}/\gamma_c; \gamma_c = 1.25 \tag{4.75}$$

The unit weight of the soil material as being permanent is *factored* by $\gamma_G = 1.0$ and thus remains unchanged, while the unfavorable life load is factored by $\gamma_Q = 1.3$ to $p_d = 26$ kPa. From a new analysis with the above parameters it can be shown that

$$\frac{E_d}{R_d} = \frac{166.4 \text{ MNm}}{277.4 \text{ MNm}} = 0.6 \le 1. \tag{4.76}$$

4.6.3 Bearing capacity calculation of single footings on stone columns

The construction of a new steel framed industrial hall with a base area of 228×144 m and a height of 12 m makes ground improvement measures necessary, since the subsoil is unsuitable for carrying the high structural loads. Below a relatively old upto 4.5-m thick layer of fill consisting primarily of cohesive material follows soft marl with depths ranging to 15 m and more, which is underlain by dense sand. The foundation of the structure consists of single footings arranged in a regular pattern of 24×12 m, with footing areas between 3.5×3.5 m and 5.8×5.8 m depending on the applied loading.

The ground improvement is achieved by arranging vibro replacement stone columns below the structure with diameters of generally 0.7 m. Below the slab, the stone columns are arranged in a square pattern of 3×3 m, and underneath the footing groups of 9, 13, and 25 stone columns are constructed, depending on footing size and load to be carried.

In the following, this project will also serve as an example to illustrate the way of assessing the bearing capacity of single footings founded on vibro stone columns in the ULS. Figure 4.31 shows the soil profile and the plan view of a highly loaded 5.8×5.8 m footing carrying a maximum design load of $V_d = 10.1$ MN. Table 4.9 summarizes the soil parameters necessary for this bearing capacity calculation. The settlement improvement

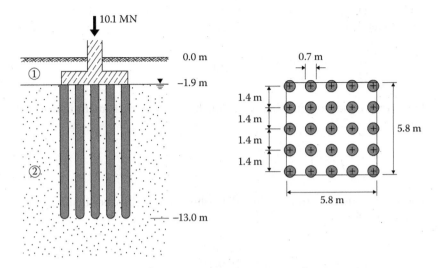

Figure 4.31 Cross section and plan view of single footing founded on 25 stone columns.

Table 4.9 Characteristic soil parameters

	Layer 1: fill	Layer 2: marl	Stone columns
Unit weight γ (kN/m³)	16	22	21
Submerged unit weight γ' (kN/m³)	6	10	11
Constrained modulus E_{oed} (MN/m²)	–	20	120
Poisson's ratio μ	–	0.45	0.33
Young's modulus E (MN/m²)	–	5.3	81
Effective friction angle φ' (°)	–	27	40
Effective cohesion c' (kN/m²)	–	10	0
Undrained shear strength c_u (kN/m²)	–	25	–
Coefficient of earth pressure at rest K_0	–	0.55	–

factor β_1 according to Priebe (1995) can be calculated using, for example, Equation 4.52:

$$\beta_1 = 2.3 \tag{4.77}$$

The assessment of the bearing capacity now follows the procedure as proposed by Barksdale and Bachus (1983) and as described in Section 4.3.3 using characteristic parameters and first calculating a characteristic bearing resistance R_k. Using partial safety factors for actions and resistances it then needs to be shown (according to Eurocode 7) that

$$V_d \leq R_d \tag{4.78}$$

The area replacement ratio can be calculated as the sum of the stone column areas A_c divided by the footing area A_F:

$$a_c = \frac{\sum A_c}{A_F} = \frac{25 \cdot [(\pi \cdot 0.7^2)/4]}{5.8^2} = 0.29 \tag{4.79}$$

The stress concentration factor n can be computed from Equation 4.12 with $\beta = \beta_1 = 2.3$:

$$n = \frac{\sigma_c}{\sigma_s} = \frac{\beta - 1}{a_c} + 1 = \frac{2.3 - 1}{0.29} + 1 = 5.5 \tag{4.80}$$

The stress concentration within the columns can thus be identified from Equation 4.6:

$$n_c = \frac{n}{1 + (n - 1) \cdot a_c} = \frac{5.5}{1 + (5.5 - 1) \cdot 0.29} = 2.4 \tag{4.81}$$

Numerical analysis of the situation showed vertical stresses in the columns close to the footing between 500 kPa and 1 MPa depending on the position underneath the footing, which fits quite well to column stresses of

$$\sigma_c = n_c \cdot \sigma = 2.4 \cdot 300 = 720 \, \text{kPa} \tag{4.82}$$

As described in Section 4.3.3, quick loading conditions are assumed with the undrained shear strength c_u prevailing in the soil ($\varphi_s = 0°$). Therefore, the composite shear strength can be computed from Equations 4.26 through 4.28:

$$\tan\varphi_{avg} = n_c \cdot a_c \cdot \tan\varphi_c = 2.4 \cdot 0.29 \cdot \tan 40° = 0.58$$
$$\Rightarrow \varphi_{avg} = 30° \tag{4.83}$$

$$c_{avg} = c_u(1 - a_c) = 25(1 - 0.29) = 18 \, \text{kPa} \tag{4.84}$$

Therefore, the inclination of the failure plane within the column group computes from Equation 4.25 to

$$\delta = 45° + \frac{\varphi_{avg}}{2} = 45° + \frac{30°}{2} = 60° \tag{4.85}$$

The square column group in the soil is approximated by a composite material of circular shape. The equivalent diameter B computes to

$$B = \sqrt{\frac{5.8^2 \cdot 4}{\pi}} = 6.5 \, \text{m} \tag{4.86}$$

According to Figures 4.18 and 4.31, the depth of the failure wedge is

$$B \cdot \tan \delta + 1.9\,\text{m} = 6.5\,\text{m} \cdot \tan 60° + 1.9\,\text{m}$$
$$= 11.3 + 1.9\,\text{m} = 13.2\,\text{m} \tag{4.87}$$

The mean horizontal stress level q is estimated from the initial stresses in the soil according to Figure 4.18:

$$q = k_0 \cdot \frac{\gamma_1 \cdot 1.9\,\text{m} + (\gamma_1 \cdot 1.9\,\text{m} + \gamma_2' \cdot 11.3\,\text{m})}{2} = 48\,\text{kPa} \tag{4.88}$$

The rigidity index I_r of the soil can be computed from the Young's modulus E, the Poisson's ratio μ, the mean stress level q, and the shear strength as

$$I_r = \frac{E}{2(1+\mu) \cdot (c' + q \cdot \tan \varphi')} \tag{4.89}$$

In case of undrained conditions, Equation 4.89 can be reduced, with $\mu = 0.5$ for deformations at constant volume, to

$$I_r = \frac{E}{3 \cdot c_u} = \frac{5300}{3 \cdot 25} \approx 70 \tag{4.90}$$

The Vesic cavity expansion factors F_c' and F_q' can be taken from Figure 4.19 or for undrained conditions ($\varphi_s = 0$) from the following equations:

$$F_c' = \ln(I_r) + 1 = 5.3$$
$$F_q' = 1.0 \tag{4.91}$$

The ultimate lateral stress in the surrounding soil can then be computed:

$$\sigma_3 = c_u \cdot F_c' + q \cdot F_q' = 25 \cdot 5.3 + 48 \cdot 1.0 = 181\,\text{kPa} \tag{4.92}$$

According to Equation 4.22 the ultimate vertical stress on the column group then amounts to

$$q_{\text{ult}} = \sigma_3 \cdot \tan^2 \delta + 2 \cdot c_{\text{avg}} \cdot \tan \delta$$
$$= 181 \cdot \tan^2 60° + 2 \cdot 18 \cdot \tan 60° = 605\,\text{kPa} \tag{4.93}$$

The characteristic bearing resistance R_k computes from the ultimate vertical stress as

$$R_k = q_{\text{ult}} \cdot A_F = 0.605\,\text{MN/m}^2 \cdot 5.8^2\,\text{m}^2 = 20.3\,\text{MN} \tag{4.94}$$

The design value of the bearing resistance R_d is then computed from

$$R_d = R_k / \gamma_{R,\nu} = 20.3\,\text{MN} / 1.4 = 14.5\,\text{MN} \tag{4.95}$$

and the design check prevails as

$$V_d = 10.1\,\text{MN} \le 14.5\,\text{MN} = R_d \qquad (4.96)$$

4.6.4 Some results of a parametric study of stone column group behavior

Field measurements to study the group behavior by investigating the interaction of load application, stone columns, and soil for variable foundation stiffness, column pattern, and length are almost prohibitive in view of the practical difficulties and costs. Alternatively, finite-element-based numerical analyses that are properly adjusted by accompanying field measurements can provide an interesting insight into vibro stone column behavior under load.

Figure 4.32 shows details of such a study for a rectangular concrete foundation on stone columns of variable number and depth. The parametric 3D study used APDL programming language for the FEM ANSYS programming system based on the finite element method (Kirsch, 2004). From the many parameters investigated, the following concentrates on the influence of the

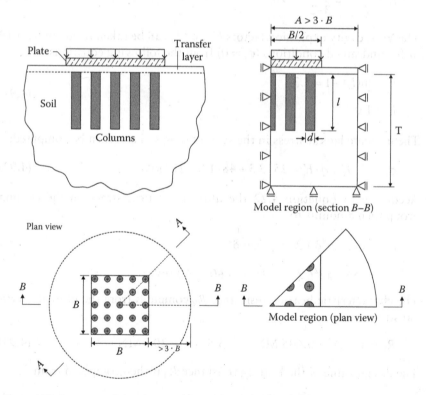

Figure 4.32 System model and symmetries used in computations.

area replacement ratio a_c (for column groups defined as the ratio between the total area of stone columns ΣA_c to the footing area A_F according to Equation 4.90), foundation stiffness, and relative column length λ (ratio of column length l and depth T of the soft soil layer according to Equation 4.91); on the improvement factor β; and on the stress concentration factor n. The study allowed computing groups of between 4 and 81 columns supporting a square foundation with a maximum width of $B = 20$ m. The friction angle φ_c of the stone column material varied between 35° and 50°.

$$a_c = \frac{\Sigma A_c}{A_F} \tag{4.97}$$

$$\lambda = \frac{l}{T} \tag{4.98}$$

When looking at the influence of the area replacement ratio a_c and the friction angle φ_c of the stone column material on the improvement factor β, Figure 4.33 shows this relationship for rigid square footings founded on a variable number of stone columns with $\lambda = 1$ and $\lambda = 0.75$. In order to obtain an improvement factor of $\beta = 2$ for a foundation with stone columns extended to the bearing stratum ($\lambda = 1$), the necessary replacement ratio is between an uneconomically high value of $a_c = 0.7$ for $\varphi_c = 35°$ and a realistic value of $a_c = 0.25$ for $\varphi_c = 50°$. The adjacent figure for $\lambda = 0.75$, whereby the columns reach only 75% of the necessary improvement depth, illustrates the importance extending the stone columns whenever possible into a competent bearing stratum, since the same configuration of stone columns with $a_c = 0.25$ achieves only improvement factors between $\beta = 1.3$ for $\varphi_c = 35°$

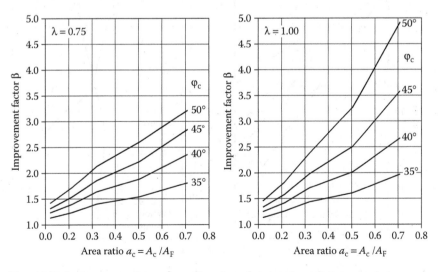

Figure 4.33 Improvement factor β as a function of replacement factor a_c and stone column friction angle φ_c for different relative column lengths λ.

and $\beta = 1.9$ for $\varphi_c = 50°$ when the column is floating. For the same a_c value, the improvement factor drops further with $\lambda = 0.5$ to $\beta = 1.3$ for $\varphi_c = 35°$ and $\beta = 1.7$ for $\varphi_c = 50°$.

The study also revealed that the improvement factor β is—within the limits—invariable for different foundation dimensions but equal foundation stiffness K_S.

$$K_S = \frac{E_F}{12 \cdot E_s}\left(\frac{H}{B}\right)^3 \frac{T}{B} \tag{4.99}$$

where:

E_F is the modulus of elasticity of footing in kPa
E_s is the modulus of soil to be improved in kPa
H is the thickness of footing in m
B is the width of footing in m
T is the thickness of soil layer to be improved in m

Figure 4.34a gives the load settlement curves for different square footings founded on stone columns with identical area replacement values $a_c = 0.20$. As expected, settlement rises with increasing load and loaded area as a result of the different vertical stresses in the ground, which are shown in

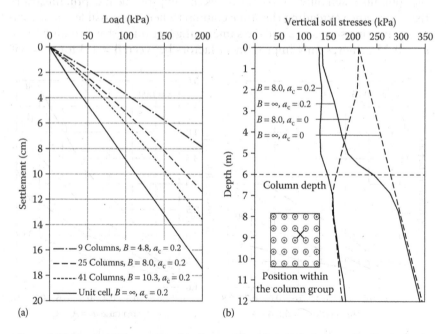

Figure 4.34 Vertical deformations (a) of square foundations and stress distribution in the soil (b) for variable dimensions and constant a_c ($T = 12$ m, $\lambda = 0.5$).

Figure 4.34b for a square footing ($B = 8$ m) founded on 25 stone columns with $a_c = 0.20$ in comparison with the infinite grid situation. Stress concentration within the stone columns results in stress relief in the soil as compared with the situation without stone columns ($a_c = 0$). Below the stone columns—here at a depth of 6.0 m—soil stresses of the improved and unimproved case are approaching each other again at about 7–8 m depth.

The distribution of stresses between column and soil, which is in the first place responsible for the soil improvement, depends both on a_c and φ_c, and also on the foundation situation of the stone column itself (founded in competent stratum with $\lambda = 1$ or floating with $\lambda < 1$). Figure 4.35 shows the situation of a square footing with $B = 8$ m founded on 25 stone columns with constant replacement ratio $a_c = 0.32$ but different lengths l between 4 and 18 m. Layer 2 has a considerably higher modulus (factor 40) than layer 1 needing improvement.

The findings of the numerical analysis are presented in Figure 4.36 giving plots of (a) the stress concentration factors $n = \sigma_c/\sigma_s$ against depth for different relative stone column lengths λ, and (b) the column stress concentration factor $n_c = \sigma_c/\sigma$.

When looking at the n_c development with depth for the square footing, it is interesting to see that column stresses rise again even in very long columns ($\lambda = 1$) beyond a relative depth of about $t/l = 0.55$ to a level of $n_c = 1$ at the toe of the column. For floating columns, column stresses decrease gradually with column depth to n_c levels of 0.8–0.6.

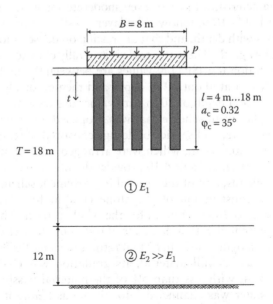

Figure 4.35 Situation considered for study of column length influence.

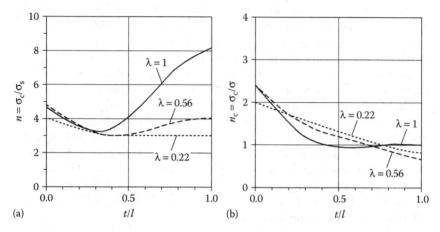

Figure 4.36 Stress concentration factors (a) *n* and (b) n_c for a rigid square foundation ($B = 8$ m) on 25 stone columns for $\sigma_c = 0.32$ and $\varphi_c = 35°$.

4.7 CASE HISTORIES

4.7.1 Wet vibro replacement stone columns for a thermal power plant

For a new 1500 MW coal-fired thermal power plant in southern India, all ancillary structures such as cooling towers, service buildings, and special storage tanks were founded on improved ground, because the original soil had insufficient characteristics to carry even moderate loads. It consists of soft clays to a depth of 7–10 m followed by layers of silty sands and silts whose density increases with depth, and that are resting on dense sand at a depth of about 20 m below grade. The water table is generally close to ground surface. Standard penetration test (SPT) *N*-values are between 0 and 2 in the upper 7 m rising to $N = 5$ at about 10 m depth and then gradually increasing to well over $N = 30$ at a depth of 20 m with values in excess of $N = 50$ below.

The vibro replacement scheme that was proposed for structures such as tanks, cooling towers, and service buildings consisted of stone columns of lengths between 7 and 10 m, which were arranged in triangular grids of 2 and 2.5 m equal spacing. Table 4.10 provides details of the structures, their dimensions and loads, and of the ground improvement scheme adopted.

In 2008, the construction of the stone column foundation was finished, amounting to 285,000 lin.m for the whole project. The vibro stone columns were executed with a diameter of 0.9 m by the wet vibro replacement method employing the Keller M vibrator and using 12/75 mm crushed stone aggregate as backfill material. Its gradation was characterized by D_{60}/D_{10} value of 4, with less than 5% of the material passing the 12 mm sieve. Process water was channeled into silt ponds before it was released into a nearby creek. Quality control was performed with a standard data

Table 4.10 Details of the structures founded on vibro stone columns

	Storage tanks	Cooling towers	Buildings
Dimensions	16.6 m diameter 11.5 m height	144 × 32 m	11 × 5 m
Loading	135 kPa	90 kPa	55 kPa
Type of foundation	Raft	Raft	Strip footings
Allowable settlement	100 mm	75 mm	40 mm
Triangular grid spacing	2.5 m	2.0 m	2.5 m
Column length	7.0 m	11.0 m	7.0 m

acquisition system measuring depth and amperage versus time of execution. In addition, numerous load tests were performed on single columns and on groups of three columns as proof of the load settlement performance of the structures. Figure 4.37 shows the stone column arrangement for a storage tank together with a cross section through its foundation.

Acceptance tests for the vibro stone columns were carried out in accordance with the Indian Standard (IS) 15284, which require tests on single columns within a group of at least seven stone columns and load tests on a group of three columns out of a total of a minimum of 12 stone columns. Footing size required by the standard follows the unit cell principle in specifying it on the basis of "the effective tributary soil area of stone column for a single column test and three times the effective area of single column for a three column group test" (IS 15284, 2003). The test footings were of circular shape, sufficiently rigid, and centrally placed over a granular blanket

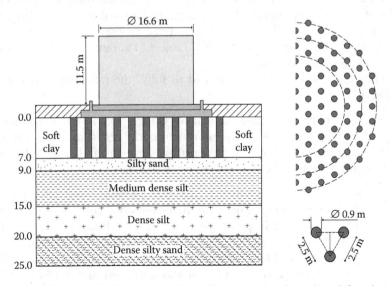

Figure 4.37 Vibro stone column layout for and cross section through tank foundation.

of a minimum thickness of 0.30 m. At 150% design load, the single column settled generally not more than 15 mm with the three column group tests settling less than 25 mm. At the chosen design load of 300 kN per stone column, the settlements were only 8 or 8.5 mm, respectively.

4.7.2 Vibro replacement soil improvement for a double track railway project

The Ipoh–Rawang and Ipoh–Padang Besar double track railway construction projects in Malaysia have been under way since 2001 and involve about 330 km of new line. The project crosses five states and encompasses the construction of a new track near, or next to, the existing railway track. Ground improvement works were necessary for about 32 km, where the new track crosses over poor soils. Some stretches of the track cut through valuable farmland. In such areas, stone column installation was restricted to the dry bottom feed method to avoid contamination of the neighboring fields by the silt- and clay-loaded process water. The wet vibro replacement method was used when low headroom conditions below existing road bridges only allowed the execution of short columns of about 5 m length. As the soil improvement work had to be carried out as close as 2.5 m from the existing live track, very stringent safety measures were put in place to minimize the risks for passing trains and for the workmen and their rigs. Vibro stone columns were also used to support earth embankments and reinforced earth walls approaching new bridges crossing the railway tracks. Figure 4.38 shows a typical situation of a dry bottom feed vibrocat operating next to the existing track.

Where ground improvement was necessary, the subsoil consisted generally of soft clay deposits to about a depth of 7 m and of firm clay to 10 m. At that elevation, stiff clay was encountered whose stiffness increased at 15 m depth and beyond considerably. Table 4.11 summarizes the soil conditions in an idealized profile.

The design of the railway line had to fulfill the following performance criteria:

- A 25 mm maximum settlement with a maximum deflection of 1:1000 of the track over 6 months from the certificate of acceptance
- A minimum factor of safety against slope instability of 1.2 during construction and of 1.4 during service stage

The vibro stone column scheme that was proposed and adopted consisted of a rectangular grid of 1.0 m diameter stone columns with lengths of generally 10 m, but in places up to 18 m. The distance between vibro probes within the grid was 2.25 or 2.5 m, resulting in area replacement factors $a_c = 0.16$ and 0.13, respectively. In total, an area of over one million square meter covering unsuitable soil was improved by the vibro replacement method by the end of 2008 using M-type Keller bottom feed depth

Figure 4.38 Vibrocat in operation next to existing railway track. (Courtesy of Keller Group plc, London, UK.)

vibrators. The key element of the quality assurance program was the standard data acquisition system recording depth and amperage over time, a conventional measurement of the stone consumption and frequent load tests on single columns. The load tests were performed with relatively small steel plates of 1.5 × 1.5 m or even only 1 m diameter as test footings. The regularity of the test result served as a good indicator for the steady execution of the stone columns and their performance under load.

In this context, it should be remembered that zone load tests, or at least tests with footing sizes equal to the unit cell area, would deliver more significant results, the latter at relatively low extra costs, to better interpret the load settlement performance of the structure.

Table 4.11 Idealized soil profile for stretches of the railway line needing ground improvement

Layer	Depth below GL (m)	Soil type	Average SPT N-value	Undrained shear strength (kPa)	Deformation modulus (kN/m²)
1	0–4	Soft clay	0–1	10	1,000
2	4–7	Soft clay	2	20	2,000
3	7–10	Firm clay	4	30	3,000
4	10–15	Stiff clay	10	60	18,000
5	>15	Very stiff clay	>15	–	–

4.7.3 Vibro replacement foundation for the new international airport at Berlin

The new international airport for Berlin is expected to go into operation in 2017. For the construction of infrastructural measures to connect the central terminal building with the highway and with high speed and regional rail traffic systems, ground improvement by vibro replacement was necessary. Subsoil at the site consists of marl and sand layers of glacial origin that are relatively soft to a depth of about 8 m below ground surface, where generally stiff marl follows, characterized by CPT cone resistance values typically well in excess of 15 MN/m². Design target of the ground improvement scheme was a reduction of the settlements deriving from the upper soft layers by a factor of 2 for single footings and a deformation modulus of 21 MN/m² below rafts and approach ramps. The necessary spacing of the vibro stone columns was determined by load tests on groups of four columns for structures and on single columns adopting the unit cell concept for area loads.

All single column load tests were performed with a test load of 200 kN/m² on square test footings of 1.25 × 1.25 m and 2 × 2 m, each supported by a 9 m long vibro stone column of 0.5 or 0.6 m diameter. Figure 4.39 shows the load test setup together with a CPT diagram which is representative of the site. Table 4.12 gives details of the load tests which were evaluated by the method described in Section 4.4.

By adopting the unit cell concept the settlement of a single column determines the equivalent deformation modulus E^* by introducing the equivalent column length l^* from Figure 4.28 into Equation 4.47. The necessary grid spacing for the target modulus can then be determined graphically, here by extrapolation according to Figure 4.40, where a necessary area A_n = 7.4 m² for a square grid of 0.6 m diameter stone columns was obtained for the required design modulus E_d = 21 MN/m². The ground improvement works were performed using conservatively a square grid of 0.6 m diameter stone columns of 9.0 m length with 2.5 m spacing.

While the necessary area replacement factor for the infinite grid was just a_c = 0.05 to achieve a modulus of 21 MN/m², the evaluation of the column group tests resulted in a required a_c = 0.09 for single footings to reduce settlements by 50%. Prior to the load tests, large shear box tests were performed in the laboratory with the stone column material anticipated for the ground improvement works. It consisted of crushed granite rock of 10–35 mm grain size. In the laboratory an average friction angle of 57° was determined for the backfill material at a density of 1.9 t/m³, safely allowing the use of column friction angles of 45° in the design calculation.

An area of about 20,000 m² was improved by stone columns on this project in 2008 using Keller bottom feed M vibrators. Quality control measures included standard data acquisition systems to record the usual parameters

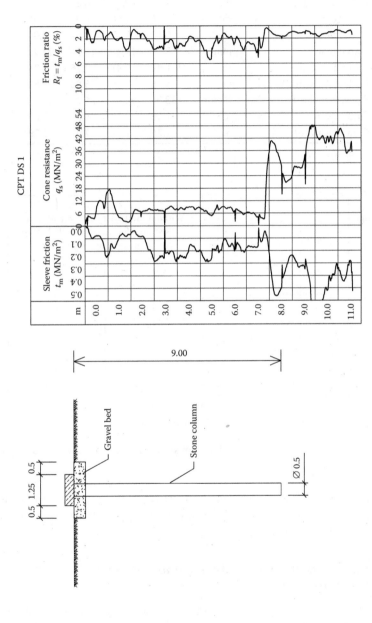

Figure 4.39 Load test on single column and representative CPT result.

Table 4.12 Evaluation of single column load tests

Test load 200 kN/m²		Test 1	Test 2	Test 3	Test 4
Column dia. d_c	m	0.6	0.5	0.6	0.5
Footing area A	m²	1.56	1.56	4.0	4.0
$a_c = A_c/A$	–	0.18	0.13	0.07	0.05
Settlement s	mm	8	11.4	15.7	23.9
Equiv. diameter d_e	m	1.41	1.41	2.26	2.26
Equiv. col. length $l*$	m	5.36	5.36	5.88	5.88
Equiv. modulus $E*$	MN/m²	133.9	94.0	75.0	49.2
Grid area A ($d_c = 0.6$ m)	m²	1.56	2.17	4.0	5.65
Grid spacing[a] ($d_c = 0.6$ m)	m	1.25	1.47	2.0	2.38

[a] For square grid.

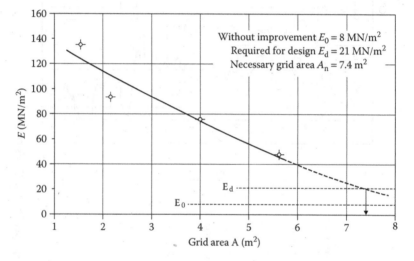

Figure 4.40 Determination of grid spacing from single column load test results.

including stone take. Early settlement measurements on the structures indicate that the performance is considerably better than expected and that the design criteria were met.

4.7.4 High replacement vibro stone columns for a port extension

The majority of vibro stone column contracts are performed at area replacements of about 5%–30%. It is feasible to provide higher replacement when higher bearing capacity, tighter settlement, or higher effective angle of friction for slope stability are required. However, the vibrator when penetrating the ground induces lateral displacement of the soil. If the stone columns are constructed too close to each other, excessive heave can be generated. This can be overcome for treatment above the water table by preboring at each vibro stone column location, and in this way area replacements of up to 50%–60% have been achieved.

Even higher replacement ratios were necessary for two port extension projects in the United Kingdom. Both involved sand fill being placed behind new quay walls where the presence of deep weak clays and silts required special design considerations. While the specified short- and long-term settlement criteria could be achieved by soil improvement using vibro stone columns at about maximum 40% area replacement, the quay wall stability required the composite soil to be virtually wholly granular in character and replacement ratios of 80% and 75% had to be performed.

The first project was based on nominal 1.2 m diameter stone columns on 1.28 m triangular grid, resulting in a replacement ratio of $a_c = 0.8$. The larger-than-normal columns were achieved with the wet method by high pressure water flushing and using Keller S300 vibrators. Intensive investigations by CPT and piezocone had permitted the development of contours for the base and surface of the soft clays and silts across the area. The scheme was then designed for the vibrator to be lowered at least 1 m deeper than the contour base, with the stone columns to be performed to at least 2 m above contour surface of the soft soils. vibro compaction was then performed to the overlying sands up to ground level.

About 50% of the soil improvement works were performed at lower levels behind the tubular pile wall between high tides. It was noted that even when water flushing was used to wash out the required soil volume, the piles were being displaced when the compaction was being applied to the stone to achieve the specified high angle of friction. A practical approach and sequence of construction combined with monitoring of the piles restricted these movements to within acceptable limits.

This project involved the construction of about 10,000 columns/compactions to depths of up to 20 m. Back analysis of the stone consumption revealed that about 98.7% of the target area replacement ratio had been achieved.

The second project was more complex since there were real stability concerns and the vibro treatment had to be performed between tie bars to even more variable soils, the extensive and thick soft clays and silts sometimes being present to immediately beneath the working level above high tide. In view of the experience of the first project two advance trials were performed. The first was to confirm that the required stone column diameter of 1.2 m and hence replacement ratio of 75% could be achieved. The second was to examine the effect of the vibro stone column construction using high power vibrators to within 1 m of the sheet pile wall with external support bund.

The results of the second trial have been reported by Slocombe and Smith (2008), where analysis of in-situ earth pressure cells suggested that K values of about 1 could be developed against the sheet piles during vibro compaction. Monitoring procedures were therefore put in place for the proposed main quay combi wall contract works, comprising large-diameter tubular piles with sheet pile infill. It was also concluded that "by installing stone columns away from the wall the peak earth pressures acting on the wall are lower than if the columns were installed towards the wall."

The works commenced with vibro treatment to the area in front and behind the proposed piled anchor wall to the proposed tie bars. The ties were installed, and the area then filled using clean sand to about high tide level. Vibro stone columns were then constructed, with sand compaction above, between the buried tie bars at two levels and at about 3.65 m centers. Every column location was very carefully set out to avoid damaging the ties and, when the tie bars were subsequently exposed for tensioning, it was confirmed that no damage had been caused.

Again, about 10,000 vibro probes (stone columns/compactions) were constructed to depths of up to about 20 m on a triangular column grid of 1.32 m. Measurement of the stone take revealed that about 97.8% of the $a_c = 0.75$ target replacement ratio had been achieved. As an illustration of the complexity of the works, the depths of treatment and elevations for the tops of the stone columns resulted in groups of no greater than six adjacent columns being of identical geometry and hence instruction.

A further phase of this port development was built where the soft clay and silt was thin and easily removed by advance dredging. The area behind the tubular pile main quay wall was then filled by placing hydraulic sand fill in about 15 m depth of water. Vibro compaction was then performed prior to the installation of tie bars using Keller S340 vibrators to achieve the specified angle of friction and settlement performance. However, the quay wall design required that the K value be limited to 0.5. Advance trials were therefore performed to confirm how far behind the wall both high- and lower-powered vibrators could be used. These revealed that better control of the wall movements could be achieved by installing the closest line of compaction using lower-power vibrators prior to using high-power vibrators further away.

4.7.5 Vibro stone columns for settlement control behind bridge abutments

In the course of the construction of an approach road to an existing highway in western Bavaria, the necessary embankments were built well in advance to allow deformations to occur before any bridge and road building was due to commence. These embankments typically have a height of up to about 6 m and rest on soft tertiary silts and clays of about 4.5–6.0 m thickness.

Dense tertiary sand, gravel, and marl form the competent bearing strata below. The silts and clays have a relatively low strength, $c_u = 15$ kN/m², and a modulus between 5 and 8 MN/m².

To reduce settlements of the embankment behind the bridge abutment, the soft silts and clays were improved by vibro stone columns over a length of about 35 m at both sides of the bridge. Settlement analysis was carried out using the method after Priebe (1995), with an area replacement value of $a_c = 0.22$ for nine stone column rows immediately behind the pile raft representing the bridge foundation, and of $a_c = 0.13$ for the remainder and below the embankment shoulders. These replacement values were achieved by 0.8 m diameter stone columns at square grid configurations with distances of 1.5 and 2.0 m, respectively. With a chosen angle of friction of $\varphi_c = 42°$ for the stone column material, modified improvement factors β were calculated as 3.28 for the 1.5 m grid and 2.17 for the 2.0 m grid, safeguarding a smooth transition between bridge and embankment.

In total, 1680 stone columns with an average length of 5.0 m were carried out on this project in 2008 using Bauer TR 17 bottom feed vibrators operated from MBF 10 and BG 12 base machines. Figure 4.41 shows such a typical setup.

Figure 4.41 Bauer TR 17 bottom feed vibrator operated from BG12 base machine. (Courtesy of Bauer, Bavaria, Germany.)

4.7.6 Ground improvement for the foundation of a petroleum tank farm in the Middle East

The tank farm area which serves as a case history is located in the Hamriya Free Zone in Sharjah/UAE and is typical for many other industrial developments in the Middle East. The subsoil below the relatively level ground surface at 2 m above sea level consists of about 2–3 m of artificial sand fill followed by slightly silty fine to medium-grained sand, which is medium dense to about 4 m; it is very loose from 4 to 6 m, and very dense below 7 m. The sand layer above this level has typically in excess of about 15% fines, which renders its suitability for compaction rather questionable. The groundwater table is at 2 m below ground surface.

The tank farm consists of 16 nos. steel tanks of the fixed roof type with diameters of 10, 15, and 20 m. The height of the tank is 18 and 24 m. The bottom of the tank rests on a 40 cm thick concrete slab with a reinforced concrete ring beam placed directly below the tank shell. In view of the deformation criteria of the tanks, the engineer proposed to improve the existing ground conditions which he found to be inadequate for the purpose accordingly. Cone penetration tests (CPTs) carried out for the project revealed the presence of loose sand between 3.50 and 6 m requiring compaction. In view of its high silt content and a friction ratio of above 0.5, its suitability for vibro compaction was only marginal. It was therefore decided to perform a field trial to define if the soils could be improved sufficiently by vibro compaction, alone or if ground improvement by vibro replacement with coarse backfill stone was necessary.

For vibro compaction, three different triangular probe grids with 2, 2.25, and 2.5 m distances were carried out, each consisting of 18 nos. vibro compaction probes of 7 m depth. A Keller S340 depth vibrator (see Table 3.1) was used employing a standard compaction procedure with 50 cm lifts and a holding time of 40 s. Backfill sand material was taken from a nearby borrow area to compensate the subsidence occurring during compaction keeping the working platform at its original level. In total 23 nos. of pre- and post compaction CPTs were carried out together with a zone load test on a 3 m square concrete foundation loaded to design load of 280 kPa in the 2.25-m grid area.

For the vibro replacement trial area, a 1.6 m triangular stone column grid was chosen consisting of 19 nos. Stone columns of 7 m length each. A Keller M1455 depth vibrator (see Table 3.1) was used employing the wet method with stone supply from ground surface. According to the consumption of the 25–100 mm crushed stone backfill measured during compaction, a column diameter of about 85 cm was achieved. Pre- and post compaction CPTs were performed in the midpoints of the column grid, and 5 days after the stone columns were completed a zone load test was carried out on a 3 m square concrete test foundation which was also loaded to 280 kPa design load.

It was not surprising to see that vibro compaction failed to achieve, even with the closest spacing, the specified minimum cone resistance of 8.0 MPa in the silty sand layer at about 4 m depth. The zone load test carried out in the

Table 4.13 Results of zone load tests in mm on 3 × 3 m test foundations at 280 kPa design bearing pressure

Phase	Test	Settlement (mm)
Precontract	On vibro compaction with 2.25 m grid	18.3
Precontract	On 4 nos. stone column	9.3
Postcontract	On 4 nos. stone column	5.8
Postcontract	On 4 nos. stone column	6.7

medium 2.25 m vibro compaction grid area resulted in 18.3 mm settlement at design bearing pressure of 280 kPa which also was found inadequate.

Consequently, vibro replacement with 85 cm diameter stone columns set out in an equilateral triangular grid of 1.60 m was chosen as foundation method. The load test result was used to predict the settlement performance of the tanks. Table 4.13 shows the results of the two load tests performed on vibro compaction and on vibro replacement stone columns before the foundation works commenced together with two further load tests carried out as quality assurance during the vibro replacement works. At design load of 280 kPa, all three load tests on stone columns show settlements of between 6 and 9 mm, whilst the load test on vibro compaction settled at 18 mm.

Based on the load tests, total maximum settlements for the centers and below the shell of the tanks were predicted with 26 and 20 mm, 23 and 18 mm, and 22 and 17 mm for the 20, 15, and 10 m diameter tanks, values which yet were to be verified during water testing of the tanks.

Figure 4.42 shows the tank farm in summer 2015 when the hydrostatic tests were performed on these storage tanks according to API 650 resulting

Figure 4.42 Tank farm in summer 2015. (Courtesy of Keller Group plc, London, UK.)

Table 4.14 Result of hydrostatic testing performed according to API 650

Tank no.	Diameter (m)	Water height (m)	Ring beam settlement (mm)
301	10	16	4.3
302	15	22	3.1
303	15	22	6.4
105	20	22	7.1
106	20	22	6.4
107	20	22	14.1

in unexpectedly low settlements. Under maximal load and after 24 h holding time, settlements of the tank perimeter ranged on average between 3 and 14 mm only for the different tanks. Table 4.14 displays details of six hydrostatic tests. It can only be assumed that the depth vibrator was able to compact the silty sand layer between 4 and 6 m below grade much better than anticipated and measured with the CPTs than originally anticipated, leading in this way to much reduced settlements.

4.7.7 Stone columns provide earthquake-resistant foundation for an electric power plant in Turkey

The project area is located at the Iskenderun Bay in the Hatay province of southeastern Turkey. The region has a long history of well-documented earthquakes. The combined cycle gas turbine power plant which was to be erected has a nominal generation capacity of 910 MW comprising of 38 structures to be built on over 30 m deep alluvial deposits overlaying competent basalt. Below a 3 m thick layer of gravel fill follow 18 m of loose liquefiable sand and silt layers, 12 m of silty clay and clay resting on basalt bedrock. The groundwater table is close to ground surface.

From a seismic study of representative earthquakes near the project area a design earthquake with a magnitude of $M = 7.5$, an equivalent number of stress cycles $N_{eq} = 39$, and a duration time of $t_d = 38$ s was developed and chosen for the design. Such an earthquake is characterized by an average shear wave velocities in excess of 500 m/s and a scaled peak particle velocity of 0.5 g.

The settlement criteria for the plant required all heavy structures to be founded on 100 and 120 cm diameter bored piles all resting on 1.5 pile diameter deep pile sockets in the basalt. All other structures were founded on 20 m deep vibro stone columns (see Figure 4.43). Vibro stone columns became necessary to protect the soil—also around the piles—against liquefaction and the corresponding lateral spread of the liquefied soil posing considerable horizontal loads on the piles necessitating otherwise uneconomically large pile diameters and reinforcement.

Thurner and Kirsch (2014) describe the various steps necessary for the foundation design of all structures of this power plant, whereas we will

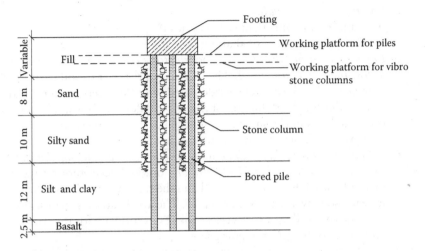

Figure 4.43 Typical soil profile and foundation for heavy structures.

restrict our considerations on estimating the safety factor against liquefaction of the sand and silt material between the large diameter bored piles during an earthquake event which would cause intolerable horizontal loads on the piles where no ground improvement would be performed.

We have seen in Sections 3.3.3 and 4.3.4 that soils losing strength during an earthquake as a result of the pore water pressure increase can be strengthened by stone columns, which provide efficient and short drainage passes for quick pore water pressure relief. We have also seen that vibro stone columns will

- Shorten the drainage path
- Increase the soil strength
- Increase the density of granular soils by the vibratory forces

The design was structured in seven steps:

1. *Determination of the geotechnical design profile.* Necessary information required is
   - SPT values (N)
   - Shear wave velocity or dynamic shear modulus (G_{dyn})
   - Unit weight (γ, γ')
   - Horizontal permeability (k_h)
   - Coefficient of compressibility (m_v)
   - Design water table
   - Strength parameters (ϕ, c)
   - Relative density (D_r)

2. *Earthquake input values.*
   - Number of equivalent stress cycles (N_{eq})
   - Relevant duration of shaking time (t_d)
3. *Improvement factor β for the chosen stone column grid is calculated according to Section 4.3.2.* The increase of the overall soil stiffness as represented by the β-value is calculated with the Priebe method and is responsible for the reduction of the vertical deformations. It is implied that it reduces in the same amount the horizontal shear stresses by an equivalent increase of the shear modulus G.
4. *Determination of layer-dependent cyclic stresses during the design earthquake using SHAKE 2000.* The original acceleration time history at bedrock level is introduced and the τ/σ'_0 ratio (CSR) is calculated for each soil layer using Equation 3.7. The acceleration is than scaled to reach design peak ground acceleration (PGA) of 0.5 g for the structures according to the design requirement.
5. *Stress ratio reduction.* The improvement factor β (here calculated by the Priebe method) is applied on the shear modulus of the unimproved soil between the piles and a reduced CSR is calculated.
6. *Number of cycles to cause liquefaction (N_l).* The N_l value is preferably established in the laboratory by undrained cyclic simple shear or triaxial tests on the in-situ soils. It can, however, alternatively, be estimated using published results as by Finn et al. (1971), provided comparability of the in-situ soil characteristics can be ensured.
7. *Reduction of liquefaction potential by improved drainage.* With the method described in Section 4.3.4 the time factor T_{ad} is calculated for the chosen drain configuration using Equation 4.45. With the earthquake severity ratio N_{eq}/N_l now being established the respective chart in Figure 4.24 can be chosen. The intersection of the d/d_e value (relative stone column spacing) represented by the drain configuration with the calculated time factor T_{ad} in the chart gives the greatest pore pressure ratio $r_g = u_g/\sigma'_v$ allowed in the design.

For the 80 cm diameter stone columns placed in square pattern of 2.1 m within the 100 cm diameter bored piles a time factor $T_{ad} = 21$ was calculated for the upper sand layers. With an earthquake severity ratio $N_{eq}/N_l = 39/11 = 3.4$ the lower right graph in Figure 4.24 is chosen where the intersection of $T_{ad} = 21$ with $d/d_e = 0.34$ gives $r_g = 0.6$. Defining the safety against liquefaction of a specific layer as

$$\eta = 1/r_g \tag{4.100}$$

We see that for the silty sand layer considered in this instance a safety factor of 1.67 against liquefaction exists. This calculation has to be carried out for all soil layers which are likely to liquefy during an earthquake event.

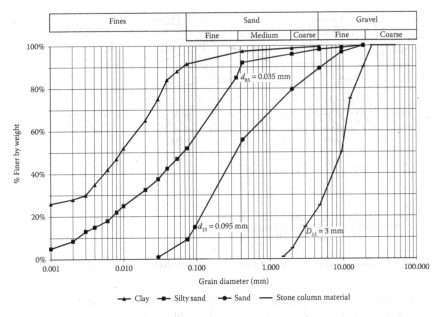

Figure 4.44 Grain size distribution of soils and stone column material.

Should the safety factor η fall below the specified value drain spacing or drain diameter have to be altered until the specified η is met. A specially composed gravel backfill with a maximal grain diameter of 25 mm was chosen for the stone column construction with its D_{15} grain diameter being established with the Saito criterion according to Equation 4.46 in order to safeguard its filter stability over time (see Figure 4.44).

The foundations for the project were completed in 2013. In addition to the 80 and 100 cm diameter bored piles, approximately 9000 nos., 20 m deep vibro stone columns with a diameter of 80 cm were carried out.

4.7.8 Seismic remediation of an earthfill dam by vibro stone columns

An instructive example of the use of stone columns to improve the safety of an earth embankment dam sealing a reservoir with a hydraulic height of 50 m and founded on alluvial deposits was published by Lawton et al. (2004) together with a companion paper by Forrest et al. (2004). The dam is situated in California and was completed in 1968. The dam as shown in Figure 4.45 was built as a zoned earthfill dam with a central clay core founded through 36.6 m of alluvial material on bedrock. Both the upstream and downstream shells with slopes of 3:1 rest on the river alluvium.

A re-evaluation of the dam's stability carried out with updated seismic hazard studies in the early 1990s revealed that the "foundation alluvium

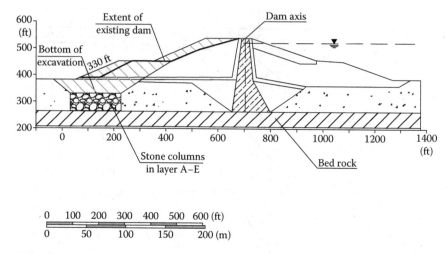

Figure 4.45 Original dam cross section and remedial alternative selected.

under the shells of the dam would liquefy and loose strength when subjected to shaking from the maximum credible earthquake (MCE)." The MCE had to be increased to a magnitude 7 and represented a significantly higher ground motion than used for the original design. Since an MCE event could lead to damage of the dam, accompanied by uncontrolled water releases and downstream flooding, the authorities responsible for the safety of the dam decided to upgrade the dam's safety to current standards.

To better understand the alternatives which were evaluated, the principal soil characteristics of the alluvial river deposits are given in Table 4.15. The above publications give details of the design, safety, and operational criteria forming the basis of the remedial design studies. Most important was that the reservoir had to remain in service during construction with a lowered pool elevation from 158.5 to 155.6 m. The following alternate approaches were actually investigated and considered:

- Construction of buttresses on one or both sides of the dam founded on the river alluvium
- Foundation improvement of the downstream buttress
- Remediation of the upstream shell of the dam
- Construction of a new dam downstream of the existing dam
- Minimum necessary construction and measures to obtain regulatory acceptance (permanent reservoir lowering)

The foundation improvement measure (b) was eventually chosen, and the vibro replacement method was selected as the most competitive from the other foundation methods which also had been studied: complete excavation to bedrock and placement of engineered fill, dynamic compaction,

Table 4.15 Principle soil characteristics of the river deposits

Alluvium layers	Elevation		Material description	Max. fines content (%)	SPT result $(N_1)_{60}$ (–)
	(ft)	(m)			
A	Above 333	Above 101.5	Sand and gravel	12	20
B	319–333	97.2–101.5	Silty sand and gravel with silt and clay lenses	30	15
C	290–319	88.4–97.2	Sand and gravel	20	19
D	280–290	85.3–88.4	Silty sand and gravel with silt and clay lenses	60	16
E	Below 280	Below 85.3	Sand and gravel	15	20

Sources: Lawton, G.M. et al., First use of stone columns in California under state regulatory jurisdiction—The seismic remediation design of Lopez Dam, in *Proceedings of Dam Safety, ASDSO's 21st Annual Conference*, Phoenix, AZ, 2004; Forrest, M. et al., Stone column construction stabilizes liquefiable foundation at Lopez Dam, in *Proceedings of Dam Safety, ASDSO's 21st Annual Conference*, Phoenix, AZ, 2004.

vibro compaction, and deep soil mixing. Figure 4.45 shows the ground improvement solution with stone columns representing the most economical and environmentally friendly foundation alternative providing the required protection against liquefaction.

Prior to construction and to better understand and evaluate the effectiveness of the vibro replacement stone columns in the existing alluvial soils, a full-scale field test was carried out. Three different triangular patterns with 2.60, 2.90, and 3.20 m probe distances were installed and tested using the SPT and Becker Penetration Test methods. As was expected the relatively high fines contents (all in excess of 12%) of the alluvial river deposits prevented significant densification with the wider spacing. Accordingly, the 2.60 m (8–1/2 ft) spacing was adopted for design in an equilateral triangular pattern. Careful measurement of the stone backfill revealed that stone column diameters varied considerably depending on the ground conditions actually encountered. It was also found that pre-densification in sandy soils would reduce column diameters, and that longer time spent with probe installation would lead to larger column diameters in silty sands. Consequently specifications were developed from the field test requiring moderately improved blow counts (minimum statistical values) in the various alluvial soils and, more importantly, stone column diameters of 100–120 cm in clean sands and sandy gravel, 165–180 cm in silty sands. Considering the relatively close column spacing, these large column diameters result in very high replacement ratios of between 38% and 45%, which were used to develop the composite strength for stability and seismic deformation analysis. The composite strength was calculated according to the method proposed by Mitchell (1981) using an internal friction angle of 40° for the stone column and 37° for the densified alluvial material between the columns. The capability of stone columns to provide also additional excess pore water dissipation was conservatively neglected in the foundation design. More interesting details can be found in the papers already mentioned above.

The block of stone columns below the downstream buttress was 76.30 m long and extended over the whole width of the dam. In total 55,000 lin.m of stone columns were installed using 81,000 ton of crushed rock resulting in an average column diameter of 100 cm.

The dry bottom feed method was used to install the stone columns to depths of 9.00–29.00 m. It is not surprising that the large column diameters together with the high replacement ratios selected for this project and the densification to be achieved in the alluvial deposits could only be accomplished using capable stone columns construction equipment (Keller S-340 vibrators) operated by experienced personnel. The responsible engineer concluded that "some level of design uncertainty in the effectiveness of *stone column* densification is unavoidable. A thorough geotechnical investigation throughout the site area can help minimize that uncertainty" (Lawton et al., 2004). In addition, the high standard QA/QC program together with a real-time data acquisition system provided at all times during construction an additional back-up security to detect zones in the foundation soil requiring different treatment than anticipated.

Chapter 5

Method variations and related processes

5.1 GENERAL

Vibratory deep compaction as described in Chapters 3 and 4 utilizes, as its main tool, a depth vibrator creating horizontal vibrations when working in the ground. An alternate method using a top vibrator and vertical excitation to compact sand was developed in the late 1960s for the U.S. market; it drove a 750 mm diameter steel pipe into the ground using a vibratory hammer. This so-called terra probe method was then, in Europe, modified by replacing the steel tube by steel H-beams and purposely built steel planks, known today as the vibro wing method or the Müller resonant compaction (MRC) method. In contrast to vibro compaction, where the depth vibrator remains in the ground generating horizontal vibrations, the vibro wing or MRC methods both utilize vertical vibrations that are transmitted from the steel planks by shear stresses into the soil with the vibrator itself remaining outside the ground (see Section 3.2.1). These methods are restricted to compaction depths of up to about 15 m. Penetration of the vibratory Y- or double-Y-shaped planks is achieved at about 25 Hz, whereas compaction is affected by considerably lower frequencies (16 Hz). To support the shearing effect in the soil, the plank has numerous apertures to support the lateral spread of the vibrations.

Process variations of the vibro replacement stone column method include both material and equipment variations. When concrete or mortar as backfill material replaces stone or gravel these columns are generally called vibro concrete columns (VCCs) or vibro mortar columns (VMCs). They are preferably used in soils needing improvement which are generally not or only marginally suitable for treatment by vibro replacement (see Table 4.3). Although, in the true sense, they do not or only marginally improve the characteristics of the soil for which they are employed, these methods are included here (see Section 5.1) in the general context of ground improvement since the methods have gained general acceptance using the same or similar equipment as with the vibro replacement method.

Other process variants use other tools than the depth vibrator to create a stone or sand column. These are, for example, the composer sand compaction pile method, the Franki gravel pile method, the controlled modulus column (CMC) method, the Geopier method, and the dynamic replacement method (see also Table 1.1).

The sand compaction pile method has been extensively used in the Far East for improving very soft marine clays, both in offshore and onshore applications to depths of about 20 m. The process uses closed-end steel pipes, with diameters between 0.5 and up to 1.6 m, that are vibrated by vibratory hammers into the ground, and the process is largely automated including sand backfilling. Area replacement ratios are generally between 0.3 and 0.5, and settlement improvement is based on experience, with an empirical stress concentration factor $n = 3$ (Tanimoto, 1973). To reduce the settlements of large diameter replacement sand columns a method has been introduced in Germany whereby a geotextile liner encases the sand (Raithel and Kempfert, 1999). Considerable vertical deformations activate the tensile forces of the geotextile before, ultimately, horizontal deformations of the column generate supporting stresses of the surrounding soil (Raithel, 1999).

In the Franki gravel pile method, which is similar to the well-known piling method, a steel pipe is driven, after a gravel plug was formed at its base, to the required depth and is then gradually withdrawn while the coarse fill material is rammed out by means of an internal drop weight.

From a multitude of similar small diameter piling methods being installed for ground improvement purposes in soft soils at relatively close spacing, the CMC method utilizes a special displacement auger that penetrates the ground to the predesigned depth essentially without spoil. Grouting of the pile shaft is carried out through the hollow auger stem. If necessary, this process can be repeated to push the grout into the soil thereby increasing contact with it to enhance skin friction at the pile perimeter. In contrast to the vibro concrete columns, end bearing of these columns is generally avoided and their design is based on the equal strain concept of soil and columns whereby column strength can be selected to keep the ratio of soil and column modulus within reasonable limits.

By contrast with the CMC method, which is used to improve soils at considerable depths, the dynamic replacement method is restricted to moderate depth ranges of about 5–7 m. With this method, a large diameter gravel column is formed in cohesive soils by hammering crushed stone into the ground in a controlled way applying a similar procedure to the well-known dynamic compaction method, but using moderate weights and drop heights. The heavy impact of the drop weight is accompanied by very severe shearing in soft cohesive soils generally without any compaction. To avoid unwanted heave and to increase column depths pre-excavation may be necessary. Generally, the method requires experience during its execution and close supervision of the development of the imprints and their filling with coarse backfill material (Varaksin, 1990; Luongo, 1992).

The Geopier method constructs impact stone columns in soft cohesive soils by driving a steel tube of approximately 0.5 m diameter into the ground using a special vibratory hammer. After excavating, coarse backfill is placed in small quantities that are compacted utilizing an internal mandrill compactor. In stiffer soils the necessary hole, generally 0.75 m in diameter, can be drilled out using standard augers. Material backfill and compaction is then performed in the same way, employing a special compactor (White et al., 2002).

It is obvious that the design approaches that are valid for vibro replacement stone columns can also be applied to all types of columns that are installed by full or partial displacement of the surrounding soil and by subsequent insertion and compaction of granular material. With similar load-carrying mechanisms, the key design parameter is the strength of the compacted coarse backfill which is controlled by the compaction energy employed. The adequate choice of the friction angle φ_c and the area replacement factor are decisive for a realistic prediction of bearing capacity and deformation of these foundation methods.

All types of columns that consist of materials whose higher strength is controlled by hardening substances such as cement or other types of hydrating binder act as piles that develop their bearing capacity from skin friction and end bearing.

5.2 VIBRO CONCRETE COLUMNS FOR FOUNDATIONS IN VERY SOFT SOILS

5.2.1 Process description

We have seen in Chapter 4 that vibro stone columns do require sufficient containing pressure from the soil which they are to improve in order to provide necessary load carrying capacity without undue settlement. The threshold expressed by the minimum undrained shear strength is generally $c_u = 5$ kPa for these soils. When in these borderline soils (see Table 4.3), particularly in the presence of natural or artificial organic material, a vibro replacement foundation would reach its limit or should not be carried out, either in view of the loads a column would have to carry or, especially, in view of the resulting deformations, the column backfill material can be replaced by concrete. When cured, this vibro concrete column, as it is called, has a stiffness which is more than 25 times larger than that of the surrounding soil.

The vibro concrete column is constructed in a similar way and using similar equipment as with the stone column method. When dry concrete is added in the same way as with the bottom feed method (see Section 4.1), a so-called premix VCC (PVCC) is built. The special premixed dry coarse concrete behaves in the same way as normal source backfill material allowing its displacement into the surrounding soil by the vibrator movement and vibrations. Column diameters range generally between 50 and 80 cm allowing a working load of about 900 kN to be used for PVCCs.

Alternatively, liquid concrete can be pumped through a pipe attached to the vibrator delivering it directly to the vibrator point when the final depth has been reached. Vibrator together with gravel or concrete pipe displaces the native soil leaving behind a more or less cylindrical hollow space which is filled with concrete in a controlled way as the vibrator is withdrawn. vibro concrete column cross sections constructed in this way are generally oval shaped with dimensions of ca. 30 × 50 cm. When the concrete is cured, these columns, or rigid inclusions as they are sometimes called, carry loads in the same manner as end bearing small diameter unreinforced concrete piles, supported by skin friction and end bearing in stiffer soils. Their bearing capacity can effectively be increased by the formation of a bulb end in gravel or stiff clay rather than by increasing their diameter. Working loads can be as high as 1200 kN (Sondermann and Wehr, 2013). Figure 5.1 shows the main working sequences when constructing a vibro concrete column. They are very slender load bearing elements, and therefore column verticality and integrity is of utmost importance.

When the stone backfill of a vibro replacement stone column is replaced by concrete for only a part of the column length, so-called hybrid columns are formed. They are particularly useful in situations where soil conditions and structural loads would allow carrying out a standard stone column foundation, except for a stratum with questionable characteristics, especially peats and other organic material, and a thickness in excess of a column diameter, a combination of the stone column method with the premix vibro concrete column technique can provide an effective solution. A normal bottom feed vibrator as described earlier will construct a standard stone column from design depth upward in the normal way by using stone or gravel as backfill until the weak soil layer is reached, where then a

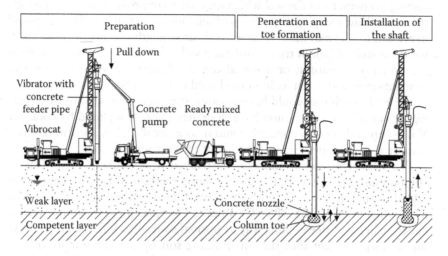

Figure 5.1 Construction sequences of vibro concrete columns. (Courtesy of Keller Group plc, London, UK.)

dry concrete mix is used in the same way bridging over the very soft layer until normal ground conditions have been reached again and normal back-fill material can be used. Needless to say that in such cases a close grid of investigation borings and soundings are necessary to clearly identify location and depth of the soft soil layer.

5.2.2 Special equipment

Figure 5.1 shows the main pieces of equipment that are used for VCC and PVCC construction:

- The vibrocat carrying the depth vibrator, which needs to be of the bottom feed type for PVCC construction, and the electric generator or the hydraulic power pack necessary for its operation.
- An air compressor and a loader to deliver the dry concrete mix.
- When VCCs are constructed the depth vibrator is modified by an attached steel pipe allowing the fresh concrete to be pumped down to the vibrator point.
- A standard concrete pump for the liquid concrete delivered to the site.

Figure 5.2 gives details of the depth vibrator modified for the construction of VCCs and PVCCs. It consists, in general, of a standard depth vibrator operating at high frequency and moderate amplitude. Attached to its outside over its full length is a small diameter steel pipe to deliver the concrete to the vibrator toe. Recent plant improvement and development aims at minimizing the cross section of the tool improving in this way the competitiveness of vibro concrete columns over other small diameter pile systems (see Figure 5.2a).

For the construction of PVCCs standard bottom feed vibrators are used (see Figure 5.2b).

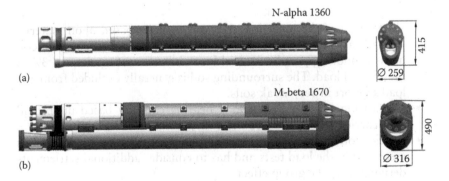

Figure 5.2 Depth vibrators modified for (a) vibro concrete column and (b) premix vibro concrete column construction and their cross sections with dimensions in millimeter. (Courtesy of Keller Group plc, London, UK.)

The data acquisition system which is indispensible for this foundation technique has to provide reliable information to the operator of the verticality of the column being constructed and, most prominently, that the withdrawal rate of the vibrator is closely controlled, coordinated, and when necessary, updated with the concrete pumping rate to safeguard continuity of the vibro concrete column shaft diameter and avoiding any column necking.

When dry premix concrete is used as backfill material for premix vibro concrete column construction, the standard dry bottom feed equipment can be used unaltered (see Figure 4.3). This method is however restricted to water bearing soils which provide sufficient moisture for the proper curing of the concrete.

5.2.3 Principal behavior and design

Both types of vibro concrete columns generally behave like unreinforced small diameter concrete piles and do not represent, in the true sense, a soil improvement technique when the displacement effect and its potential benefit on the properties of the surrounding soil are disregarded.

Vibro concrete columns using liquid concrete for their construction are generally slender and behave like unreinforced small diameter concrete piles. As they do receive only limited horizontal support from the surrounding soft soil they cannot be regarded as displacement or bored piles (see DGGT, in press).

Application of these load bearing elements (VCC and PVCC) is in certain countries controlled by special governmental permissions (Germany) or follows local codes of practice, such as ASIRI—Amélioration des sols par les inclusions rigides (see IREX, 2013) in France. In Germany their application is however restricted to soils with a minimum strength of $c_u = 15$ kPa, unless special improvement measures have been undertaken. Working loads, relying on a large number of load test results, range up to 900 and 1200 kN, respectively.

Generally, vibro concrete columns act only together with others in column groups. They are particularly suited beneath large loaded areas. Column design encompasses

- Verification of the inner load bearing capacity based upon material strength as for unreinforced concrete elements. Vibro concrete columns cannot support horizontal loads in excess of a tolerated 3% of the vertical load. The surrounding soil is generally excluded from any load support in very weak soils.
- Verification of the external load bearing capacity based upon load tests carried out on-site or on tests in similar soil conditions.
- Verification of the serviceability limit state uses the deformations measured in the load tests and has to consider additional settlements deriving from the group effect.

Priebe (2003) has extended his method to calculate settlements of stone columns in very soft soils and allows also to consider the case where the load on the column is remarkably lower than its inner strength, a case where in addition the column module is very much higher than the module of the surrounding soil ($E_c \gg E_s$). In a two-step calculation, the settlement of a group of columns can be estimated including any additional deformation below the column tow, stemming from, what is generally called, the punching effect.

The displacement effect of the bottom feed vibrator during PVCC and VCC installation and its beneficial influence on the soil properties has been investigated by Sondermann and Kirsch (2009). They report of closer bonding between column surface and soil and enhanced skin friction leading to higher bearing capacities of the columns (see also Section 4.3.5).

The effective transfer of the building loads into the column heads requires a load distribution and transfer layer of sufficient strength, which can be achieved either by a concrete slab capable of carrying the single column loads, or by a granular cushion of appropriate thickness and friction value of its material to allow an arching effect to develop according to Figure 4.12. To reduce the thickness of such a load distribution layer, a geotextile of sufficient strength may be placed across the column heads (Kempfert, 1995; Sondermann and Jebe, 1996; Topolnicki, 1996).

Preparation of the column heads for the following construction work requires special attention and care. When cutting cured VCCs and PVCCs to foundation level, cracking of the columns must be avoided and the cutting tools have to be selected accordingly. It is therefore recommended to achieve this by using handwork or only very light excavators for cutting of the columns which have not yet been cured. After careful removal of the debris, the excavations should be backfilled again for column head protection to ground level with sand or other appropriate material.

5.2.4 Quality control and testing

Since vibro concrete columns are generally quasi by definition executed for the improvement of very soft soils (although these soils are generally only bypassed and not improved in their characteristics), an adequate quality assurance program is indispensible (see also DGGT, in press). It includes all the elements which are deemed necessary for vibro replacement stone columns according to Section 4.4. In addition the verticality of the columns is very important and must be carefully controlled, measured, and reported for each vibro concrete column. In this regard, a competent, horizontal working platform is required for a stable positioning of the heavy vibrocat machine. When fluid ready-mix concrete is used, pumping pressure and rate must be carefully coordinated with the rate of extracting the vibrator from the ground and with the hollow volume it theoretically leaves behind, and which needs to be refilled without delay by the concrete. Only by this way a continuous vibro concrete column can be constructed with a stable minimum diameter.

Vibro concrete columns displace the soil according to the volume of the construction apparatus. This results in a certain amount of heave at ground level but also extends horizontally and may, when the distance to neighboring columns is too close, result in unwanted horizontal forces on already cured columns or horizontal deformations of the cylindrical shaft of fresh or curing columns. The sequence of column construction and the minimal allowable distance between columns need therefore careful consideration. As a rule of thumb, the minimum column distance should not fall below three column diameters. When no experience exists from column executions in similar soils, a trial execution in a test field is recommended where safe operational parameters can be developed.

Of equal importance is the avoidance of deformations of the column heads deriving from the heavy loads of the construction equipment (vibrocat, concrete pump, and heavy trucks). Figure 5.3 shows the damage which can occur through such circumstances and which can only be remedied at large expense.

In the presence of aggressive groundwater, the concrete for column construction has to be chosen in the same way as for a piling foundation under the same conditions. Quality assurance during construction should include sampling of the concrete used and retaining the samples for laboratory testing in the same way as for concrete piles. When necessary, cores from cured columns may also be extracted and tested in special circumstances. Column length and verticality have to be recorded.

Column integrity of vibro concrete columns, which are constructed by using liquid concrete, is particularly important and may be tested by the low-strain integrity test to detect cracks and diameter irregularities of the column shaft and zones of insufficient concrete strength. In view of the inherent variations of these columns with larger diameters in zones of extreme softness, interpretation requires experience. Unexplainable test result may require the column to be excavated or tested by core drilling.

Figure 5.3 Damage to vibro concrete column heads caused by uncontrolled heavy site traffic. (Courtesy of Keller Group plc, London, UK.)

For verification of the load bearing capacity of single columns or column groups, standard static load tests are necessary. Dynamic load tests or high strain tests may only be carried out on vibro concrete columns whose column geometry has already been established and when sufficiently calibrated against static load tests. It needs to be ensured that the columns can bear the tensile stresses developing during the impact of the dynamic load test. For this reason, column heads of test piles can be safeguarded by a special steel reinforcement. Alternatively, the dynamic impact of the high strain test can artificially be elongated by packages of large steel springs ensuring that no detrimental tensile stresses can develop which otherwise result during stress wave propagation.

5.2.5 Suitable soils and method limitations

We have already seen that the fresh concrete of the column shaft requires lateral support from the surrounding soil during the curing process. This is particularly important in soils with thixotropic characteristics which temporarily lose their strength during installation as a result of the vibrator motion. In addition, influences from the construction of adjacent columns acting on each other during installation cannot easily be excluded. Consequently, a minimum column distance and a threshold for the minimum soil strength need to be defined and observed for the safe construction of these columns. In soils which are prone to creep which may result in negative skin friction and additional vertical forces on the vibro concrete columns, special care is necessary when using this foundation method.

When extensive heave develops during vibro concrete column construction, which cannot be controlled by changing the column pattern and which may have a negative impact on the column integrity; partial or complete pre-drilling at the column location can often provide remedy to the situation at a relatively low cost.

5.2.6 Case history: Foundation on vibro concrete columns in soft alluvial soils

In the vicinity of an existing hospital in southern Germany, a new radiotherapy center was to be built in early 2015 consisting of three additional one-storey structures measuring 19 × 18 m. One of the buildings encompassed the radiation treatment room, a heavy concrete structure, 6 m high, and with wall and roof thicknesses of 1.75 and 3.00 m, respectively.

Site investigations revealed the presence of a soft alluvial layer to about 8 m depth consisting primarily of sandy silt overlaying dense quaternary sandy gravel to about 12 m depth, where tertiary very stiff to hard silts and clays were found. Table 5.1 provides average soil characteristics established in site and laboratory tests. These show that the superficial soft layer is unsuitable to carry foundation loads without excessive settlements, which had been estimated to 4 cm for the light radiation buildings and to 9 cm for the heavy radiation room.

Table 5.1 Characteristic soil properties

Soil layer and depth	Shear parameters		Stiffness modulus
	c_u (kN/m²)	φ' (°)	E_s (MN/m²)
Superficial alluvial layer (0–8 m)			
Sandy	—	30	7
Gravelly	—	32.5	15
Silty	30	25	5
Quaternary gravel	—	35	100
Tertiary layers			
Very stiff	70	22.5	10
Hard	100	22.5	20

These deformations were incompatible for the envisaged structures, and it was decided to carry out a vibro concrete column foundation in accordance with existing German regulatory requirements. The vibro concrete columns were constructed from a stable working platform using the small diameter Keller N-Alpha vibrator (see Table 3.1 and Figure 5.2b). The minimum column diameter, depending on the vibrator diameter actually used, was chosen as 40 cm and the diameter of the column toe 80 cm. Based upon experience from the construction of vibro concrete columns in similar soil conditions, the safe characteristic vertical load capacity of the columns was chosen as 710 kN and column spacing was selected accordingly. In column construction, the column toes had to be safeguarded by enhancing the diameters to at least 80 cm and that all were actually resting on the dense gravel layer below 8 m depth with envisaged settlements below 1.5 cm.

In total, 103 working vibro concrete columns were necessary for the foundation of the medical center. To verify the selected safe load bearing capacity of the vibro concrete columns, six test columns were installed and load tested employing the high strain method and analyzed using the CAPWAP procedure (Rausche et al., 1985). The test columns had a nominal diameter of 40 cm and lengths between 5.2 and 9.1 m, respectively. Table 5.2 gives the

Table 5.2 Load tests on vibro concrete test columns (CAPWAP test results)

Column no.	Column		Activated static bearing capacity (kN)	Skin friction (kN)	Point bearing (kN)	Settlement at test blow (mm)
	Diameter (cm)	Length (m)				
1	40	6.2	1238	424	814	8
2	40	9.0	2204	1104	1100	8
3	40	5.2	834	275	559	14
4	40	6.6	1528	416	1112	8
5	40	8.8	2087	852	1235	8
6	40	9.1	2287	1025	1262	8

results of the load tests, which were carried out after the column concrete had reached a maturity of at least 15 days.

All dynamic load tests, even on the short column of just 5.2 m length, performed well and showed bearing capacities well in excess of the characteristic load chosen for the design. It should however be remembered that the execution of vibro concrete columns requires generally standard static load testing to establish their safe load bearing capacity for every project site with potential additional dynamic load testing to act as verification procedure only.

Chapter 6

Environmental considerations

6.1 GENERAL REMARKS

When not appropriately applied and executed, deep vibratory methods, vibro compaction, vibro replacement, stone column, and vibro concrete columns are potential sources of annoyance and nuisance to people and environment. They may also cause damage to adjacent structures if certain rules are ignored.

It should be appreciated that the vibro methods generate considerably lower greenhouse gases, notably carbon dioxide, than the common conventional foundation solutions, and have, with the exception of vibro concrete columns, no detrimental impact on the groundwater. Cements and other cementitious materials are commonly used in vibro concrete column construction are very caustic and need to be included in the risk assessment and enforcement of safe working conditions on-site. The general application of the vibro concrete column method for a specific project to be executed in water-bearing soils follows the same rules of applications as any concrete pile foundation in such circumstances, and the choice of specific cements may have to be observed. As we have seen previously, vibro compaction and vibro replacement use only flushing water and compressed air and, when stone columns are built, inert back-fill material (sand, gravel, or crushed stone). Should in specific cases the use of recycled concrete material pose economical advantages, not only the abrasion resistance and hardness of such materials need to be sufficiently high, but also their chemistry, when exposed to the groundwater should be reviewed to prevent unwanted groundwater pollution or any negative impact of the groundwater on the material chosen (see also Section 4.3.5).

In the majority of applications, the vibro compaction and vibro replacement methods have, besides any economical consideration, considerable environmental advantages over other foundation methods.

6.2 NOISE EMISSION

Most countries have issued national laws or regulations for noise emission of construction sites to protect residents living in the vicinity—for example, in the UK BS 5228, and in Germany DIN EN ISO 3744. These regulations generally contain permitted values of noise levels for different types of neighborhood: commercial and industrial areas, housing areas, and hospitals and nursing homes. Allowable sound levels are generally specified in dB(A) and are based on the well-established fact that human mental and physical well being is negatively influenced if continuous sound levels are in excess of (Maurer, 2008b):

- 50–55 dB(A) outside buildings during the day
- 35–45 dB(A) outside buildings by night
- 30–35 dB(A) inside buildings during the day
- 25–30 dB(A) inside buildings by night

The perception of the noise intensity, generally expressed by its rate of decay with distance ΔL, measured in dB(A), is proportional to the logarithm of the distance ratio r_1/r_2 according to

$$\Delta L = 20 \cdot \log\left(\frac{r_1}{r_2}\right) \tag{6.1}$$

With r_1 and r_2 being the distances from the source of the noise, any doubling of the distance results in a sound level reduction of 6 dB(A).

Numerous noise level measurements conducted with the typical vibro replacement equipment, consisting of vibrocat, depth vibrator, generator, and air compressor including the pay loader for the gravel backfill, indicate, at the source ($r_1 = 1$ m), a sound level of about 85 dB(A), which drops to about 60 dB(A) at a distance of 100 m. During filling of the gravel container, noise levels may rise by about 10 dB(A) for a relatively short period of time. Sound absorbing elements around diesel engines and inside gravel containers can reduce noise levels at the source by about 5 dB(A), rendering the vibro replacement method a relatively noise-nuisance-free ground engineering construction method (see Figure 6.1).

When vibro compaction work is carried out, typically using crawler cranes as base machines and water as flushing medium, the noise levels at the source will be considerably below those measured for the dry vibro replacement method.

Predicted noise level graphs such as given in Figure 6.1 allow engineers to make a prediction and appraisal of the acceptability of a chosen foundation method on the projected foundation site. In particular cases, the direct measurement and review of noise levels and their propagation

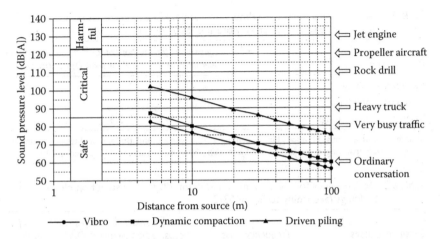

Figure 6.1 Noise levels for different ground engineering construction methods. (From O'Hara, V., Vibration and noise levels on ground improvement sites, Guidance Notes for Engineers [unpublished], Keller Ground Engineering, 2003.)

into the neighborhood may become necessary and will help increasing acceptability of a valued foundation engineering solution (O'Hara and Davison, 2004).

6.3 VIBRATION NUISANCE AND POTENTIAL DAMAGES TO ADJACENT STRUCTURES

Vibration nuisance is generally measured by the peak particle velocity (PPV) of the shear waves emanating from the depth vibrator where potential damage to structures in the vicinity of the compaction works is concerned. Such structures can be normal buildings at ground surface but also infrastructure facilities below ground. As we saw in Section 3.2.1 these vibrations may also have a direct influence on the density of granular, noncohesive soils depending on the generated ground acceleration, which can cause settlement and subsequent damage to the structure. Loose sands below groundwater are particularly sensitive to this effect if a critical acceleration of 0.3–0.5 g is exceeded.

For a steady-state vibration with a frequency f, PPV and peak particle acceleration (PPA) are connected with each other by the following expression:

$$PPA = 2\pi \cdot f \cdot PPV \qquad (6.2)$$

Different national codes and regulations, such as the German DIN 4150-3 and the Swiss SN 640312a, provide maximum horizontal or resultant PPV values, respectively, for different types of structures (Tables 6.1 and 6.2).

Table 6.1 Design values for the maximum horizontal PPV at the top floor level of a building due to steady-state vibration (according to DIN 4150-3)

Building type	Design values for the horizontal PPV of construction elements at the top floor level, DPPV$_h$ (mm/s)
Industrial buildings	10
Residential buildings	5
Very sensitive buildings	2.5

Source: From Achmus, M. et al., Untersuchung zu Bauwerks—und Bodenerschütterungen infolge Tiefenrüttlung, in *3. Hans Lorenz Symposium*, Grundbauinstitut H. 41, TU Berlin, 2007.

Table 6.2 Design values for the maximum resultant PPV for construction elements of buildings (according to SN 640312a)

Sensitivity classes	Frequency class	Design values for the resultant PPV of construction elements, DPPV$_{res}$ (mm/s)		
		$f < 30$ Hz	$f = 30–60$ Hz	$f > 60$ Hz
Normal sensitivity (e.g., usual residential buildings, office buildings)	Occasional	15	20	30
	Frequent	6	8	12
	Permanent	3	4	6
Little sensitivity (e.g., industrial buildings)		Up to two times the respective values for normally sensitive buildings		
Increased sensitivity (e.g., new residential and historic buildings)		Between 100% and 50% of the respective values for normally sensitive buildings		

Source: From Achmus, M. et al., Untersuchung zu Bauwerks—und Bodenerschütterungen infolge Tiefenrüttlung, in *3. Hans Lorenz Symposium*, Grundbauinstitut H. 41, TU Berlin, 2007.

In a recent study, Achmus et al. (2007) have investigated the impact of various depth vibrators on adjacent buildings by evaluating *PPV* measurements. In total, over 200 vibration measurements on over 20 different construction sites using four different Keller depth vibrators were analyzed. Vibrator data were according to Table 3.1 and a typical ground PPV (PPV_G) relationship versus distance from the vibrator is given in Figure 6.2 for the M- and S-vibrators, which are characterized by the nominal vibratory energy E according to Equation 6.3 and PPV_G from Equation 6.4. The factor K in Equation 6.4 is an empirical correlation factor developed from a multitude of vibration measurements of different ground engineering methods and can be found in Vrettos (2009).

$$E = \frac{W}{f} \quad \text{in kNm} \tag{6.3}$$

where:
f is the vibrator frequency in Hz
W is the nominal vibrator power in kW

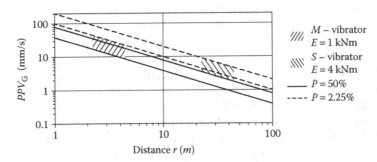

Figure 6.2 Ground *PPV* values induced by Keller M- and S-vibrators for 50% and 2.25% probabilities. (Redrawn from Achmus, M. et al., Untersuchung zu Bauwerks—und Bodenerschütterungen infolge Tiefenrüttlung, in *3. Hans Lorenz Symposium*, Grundbauinstitut H. 41, TU Berlin, Germany, 2007.)

$$PPV_G = \frac{K\sqrt{E}}{r} \quad \text{in mm/s} \tag{6.4}$$

where:
   K is the factor evaluated by field vibration measurements
   r is the distance from vibrator in m

From Figure 6.2, for a building at a distance of 6 m from the vibrator position of the S-vibrator, we get a PPV_G of 13 mm/s and for the M-vibrator 6 mm/s, when assuming a 50% (exceeding) probability. The corresponding ground PPAs (PPA_G) follow from Equation 6.2 with 2.5 m/s² for the 30 Hz S-vibrator and 1.9 m/s² for the M-vibrator, operating at 50 Hz, both values being well below the 0.3–0.5 g threshold. When considering, more appropriately, a probability of 2.25% (exceeding) only, PPA values rise to 6.6 m/s² for the S-vibrator and to 5.3 m/s² for the M-vibrator, which both are in excess of the above limit. Should the structure be founded on loose to medium dense, saturated sand, the resulting accelerations are marginally within acceptable limits for the M-vibrator at the 6 m distance but are not tolerable should the S-vibrator be used, since it could result in intolerable induced settlements of the structure. Based upon a tolerable ground PPA value of 0.5 g, the necessary distance from the building, not to cause vibration-induced settlements to it, should be about 8 m for this more powerful vibrator. When the influence of other types of depth vibrators are concerned, their vibratory energy E can be calculated using the data from Table 3.1 allowing a sensible comparison with the results as given in Figure 6.2.

   Achmus et al. have developed from these data a risk assessment method for potential building damage resulting from the impact of the vibratory deep compaction.

In a first step (Table 6.3), the *PPV* values can be predicted for the average (50%) and the worst case (2.25%) using the appropriate vibrator energy E and distance r. With the resultant ground $PPV_{G,res}$ the $PPA_{G,res}$ follow using Equation 6.2. In a second step, the propagation of the vibrations within the building need to be estimated. Neglecting unlikely resonance effect, the maximum vertical *PPV* for floor slabs can be roughly estimated by applying a transfer coefficient TC of 2 to the vertical foundation *PPV*. The horizontal *PPV* at the top floor level can be assumed equal to the horizontal *PPV* at foundation level. The third step compares *PPV*s so established with design values from relevant regulation, such as given in Tables 6.1 and 6.2. Whenever worst case values are below design values, no building damage is likely to occur. However, if predicted values, particularly the average values, exceed the design values, caution is necessary. Direct vibration measurements should be carried out, which could necessitate increasing the distance from the closest vibro probe, or alternatively usage of a less

Table 6.3 Risk assessment method for potential building damages due to vibratory deep compaction

Step	Description/equation
I	Prediction of foundation and ground vibration intensities:
	Foundation, horizontal:
	$\quad$ For $P = 50\%$: $PPV_{F,h} = 10.3(E^{1/2})/r$
	$\quad$ For $P = 2.25\%$: $PPV_{F,h} = 23.1(E^{1/2})/r$
	Foundation, vertical:
	$\quad$ For $P = 50\%$: $PPV_{F,z} = 7.3(E^{1/2})/r$
	$\quad$ For $P = 2.25\%$: $PPV_{F,z} = 17.0(E^{1/2})/r$
	Foundation, resultant:
	$\quad PPV_{F,res} = \left(PPV_{F,z}^2 + 2PPV_{F,h}^2 \right)^{1/2}$
	Ground, resultant:
	$\quad$ For $P = 50\%$: $PPV_{G,res} = 37.2(E^{1/2})/r$
	$\quad$ For $P = 2.25\%$: $PPV_{G,res} = 95.2(E^{1/2})/r$
	$\quad PPA_{G,res} = 2\pi f\, PPV_{G,res}$
2	Transfer coefficient $TC = PPV/PPV_F$:
	For horizontal vibrations of walls and floors:
	$\quad TC_h = 1.0$
	For vertical vibrations of floors and slabs, if no resonance is to be expected:
	$\quad TC_z = 2$
	$\quad PPV_h = TC_h\, PPV_{F,h}$
	$\quad PPV_z = TC_z\, PPV_{F,z}$
3	Comparison with design values/assessment:
	$\quad PPV_{F,res} \leq DPPV_{F,res}$ for $P = 2.25\%$ and $P = 50\%$
	$\quad PPV_h \leq DPPV_h$ for $P = 2.25\%$ and $P = 50\%$
	$\quad PPV_z \leq DPPV_z$ for $P = 2.25\%$ and $P = 50\%$
	$\quad PPA_{G,res} \leq DPPA_{G,res}$ for $P = 2.25\%$ and $P = 50\%$

Source: Achmus, M. et al., Building vibrations due to deep vibro processes, in *7th Conference on Ground Improvement Techniques*, Seoul, South Korea, 2010.

powerful depth vibrator or a modification of the foundation method would have to be considered. Vibration measurement can, by all means, serve as preservation of evidence in order to prove the admissibility of vibrations.

In Figure 6.3 results of such direct vibration measurements are shown from a project area in Berlin where vibro compaction was to be carried out in loose-to-medium dense medium sand for the foundation of a housing project. Buildings and commercial structures situated at the perimeter of the foundation site were potentially exposed to the vibrations emanating from the construction work. Vibro compaction was to be carried out with two different depth vibrators operating at frequencies of 49.5 Hz (Type A)

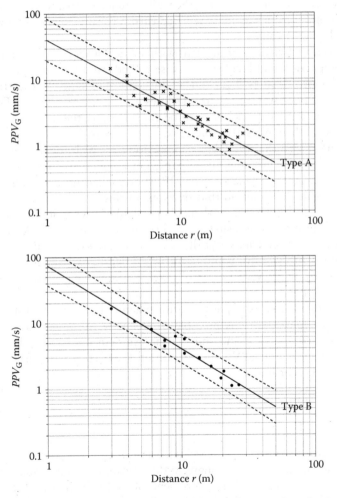

Figure 6.3 Vibration measurement (PPV) in medium sand exposed to depth vibrator vibrations. (Courtesy of GuD, Berlin, Germany.)

and 48.5 Hz (Type B) with a centrifugal force of about 220 kN creating a maximal double amplitude of 9 mm at their tips. Three test campaigns were carried out on-site each consisting of three vibro compaction probes of 7 m depth with five measuring points being placed 50 cm below ground surface and with distances from the vibrator between 3 and 26.5 m. As can be seen on the graph, the 5 mm/s *PPV* threshold for residential buildings was on average met at about 7–9 m distance from the vibrator.

When vibro compaction works are performed in granular soil behind retaining structures or inside cofferdams, to make use of the high friction angle attained from increased density, the design has to take into account different loading conditions. Initially, the wall has to support a high earth pressure that derives from a lower density but considerably higher earth pressure coefficient. Ultimately, the wall is subjected to a reduced earth pressure resulting from the much reduced earth pressure coefficient and in spite of the higher density.

In the direct vicinity of the wall with a vibro probe, the soil liquefies during compaction and local stresses rise to hydrostatic pressures with densities of approximately 2.25 ton/m^3 for saturated sand. Stress attenuation follows as pore water pressure declines (Dücker, 1957, 1968). These temporary local stress peaks have to be considered in the design, particularly where steel sheet piles are concerned.

We have seen that the vibro compaction process leads to a densification of granular soils, which is accompanied by a volume reduction manifesting itself in a ground surface lowering. Whenever adjacent structures are within reach of this deformation that develops during compaction, the structure may undergo intolerable settlements particularly when its foundations are situated close to ground surface resting on loose sandy soil. Although such foundation settlements generally do not occur, except within a distance from the depth vibrator equal to the probe depth, care should be exercised, particularly when very powerful machines are being used (cf. e.g., DIN 4150-3, Appendix C).

From the above example in Berlin where vibrations at ground surface have been measured it is possible to calculate the minimum distance at which no compaction and thus surface settlements have to be expected. With the measured PPVs, the maximum shear strain amplitudes γ_{max} are calculated according to:

$$\gamma_{max} = \frac{PPV}{c_S} \tag{6.5}$$

with measured shear wave velocities between $c_S = 104$ and 118 m/s for the different measurement campaigns in the prevailing granular soils. Figure 6.4 shows the results of these measurements accordingly. It can be seen that the lower volumetric strain level of $\gamma = 5 \cdot 10^{-5}$ below which no induced compaction of the granular soil occurs (Vucetic, 1994) is reached at a distance of

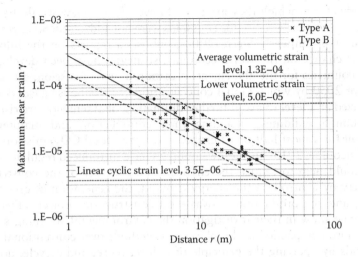

Figure 6.4 Distance from the depth vibrator versus maximum shear strain. (Courtesy of GuD, Berlin, Germany.)

approximately 4.5 m from the vibratory probe for the average of the measurements for both vibrator types. The 2.25% exceeding probability value is reached at approximately 7.5 m.

In general, deformation can be minimized by performing the compaction work in the direction toward the structure concerned, and by adopting an effective control of the material backfill during vibro probe execution or, more effectively, by keeping a safe distance from the nearest vibro probe equal to its penetration depth. Whatever the protective measures, sensitive structures must be well monitored, with regard to induced vibration and settlement throughout the vibro compaction works carried out in their vicinity.

To avoid unwanted horizontal loads developing during dry stone column installation in cohesive soils as a result of soil displacement, pre-boring is recommended to partial or full depth at the column locations close to sensitive structures. It is also recommended to start with the nearest row of stone columns and to work away from the structure needing protection. Generally not more than three rows of stone columns should be carried out in this way until normal production can be resumed again at larger distances.

6.4 CARBON DIOXIDE EMISSION

The environmental advantages of the deep vibratory ground improvement methods are evident when looking at the nature and amount of the materials used in the processes and their neutral impact on the ground and the groundwater in which they are used. However, this view would be short-sighted if their carbon dioxide emissions were not addressed. Since researchers, most prominently Stern (2006), published reports on the impact of climate change on the world economy, strategies have been developed in some countries for the abatement of carbon dioxide emission, setting targets for 2050 by which global concentration of greenhouse gases could be stabilized at a level to avoid further dramatic global temperature rises.

Although buildings, including their construction and maintenance, account for almost half of a European country's total CO_2 emission (Egan and Slocombe, 2010) only a few countries have introduced regulatory rules that could help to reduce the environmental impact of the construction industry by controlling the energy demand, the generation of greenhouse gases, and the production of waste on construction projects. Egan and Slocombe show in their investigation the environmental advantages of the deep vibro compaction and replacement methods over conventional piling methods by applying the principle of reduce, reuse, and recycle, not only qualitatively to a number of real construction projects but quantitatively by calculating the embodied CO_2 for different foundation solutions.

The comparison was made on a like-for-like basis for ground conditions that allowed both the execution of a standard piling solution and a vibro replacement stone column foundation with specifications for total settlements of 10 mm for piling and between 15 and 30 mm for the stone column alternative. The embodied CO_2 was calculated with values given in Table 6.4 which followed the principles of PAS2050 (specification for the assessment of the lifecycle greenhouse gas emissions of goods and services), which is consistent with the approach described in EN ISO 14040.

Egan and Slocombe conclude that ground improvement foundation alternatives (vibro replacement and dynamic compaction—and vibro compaction can be added, although not specifically mentioned) account for a CO_2 saving of about 90% compared with conventional piling. Typically, the carbon footprint of a normal piling project is made up of 67% embodied CO_2

Table 6.4 Embodied CO_2 for construction materials and services

Materials and services	Embodied CO_2	Remarks
Concrete (300 kg/m³ cement)	225 kg/m³	UK data average
Concrete (340 kg/m³ cement)	255 kg/m³	The Concrete Centre
Reinforcing steel	420 kg/ton	Recycled steel
Stone aggregate (quarried)	8.0 kg/ton	
Stone aggregate (recycled)	3.7 kg/ton	
Stone aggregate (virgin)	5.0 kg/ton	
Diesel fuel	2.6 kg/ton	
20 ton truck	4.4 kg/km	

Source: Egan, D. and Slocombe, B., *Proc. Inst. Civil Eng.*, 163(1), 63, 2010.

for the concrete, of 27% for the reinforcing steel and of 6% for the works execution and other supporting services. This picture changes completely for a typical vibro stone column site for which 68% of the embodied CO_2 is consumed by the stone backfill material and 32% by the works execution and transport, all this however on a much reduced overall total CO_2 emission.

Although valid European law and regulations already allow awarding public work based upon comprehensive standard of values, that is, including the cost of CO_2 emissions, and although the Stern review proposes valuing CO_2 emission at 85 EUR/t, in reality public submissions in construction continue to be assessed solely on the basis of traditional economical advantages of cost and time. Within the time horizon of the Stern review to 2050, it is not yet too late to educate decision makers in construction of the advantages of CO_2 valuation in general and of the environmental benefits which can be gained when adopting the deep vibro compaction and replacement ground improvement methods in particular.

In 2013, the European Federation of Foundation Contractors (EFFC) and the Deep Foundation Institute (DFI) published the EFFC, DFI Carbon Calculator which they had jointly developed. It is a carbon accounting methodology and provides useful carbon footprint analyses by calculating the emissions of foundation construction projects in a consistent and comparable way across the industry (EFFC–DFI, 2013). It gives also an interesting and brief description of the main carbon accounting standards developed worldwide around 2000. This tool integrates also the best practices identified in already existing carbon footprint calculators which had previously been proposed by some members of EFFC and DFI. Vibro compaction and vibro stone columns are included in a total of 15 different deep foundation and ground improvement methods.

We have seen that vibro compaction in its ideal form does not require any raw material for its execution, and that vibro replacement uses only stone aggregate as backfill material, in this way significantly reducing greenhouse gas emission. Both ground improvement methods do not leave behind troublesome obstructions in the ground encouraging in this way the unhindered reuse of the building site after demolishing of the structures. We have also seen that recycled aggregate can be used as backfill material for the stone columns provided that their physical and chemical properties are assessed properly for the purpose.

Chapter 7

Contractual matters

Traditionally, the employer or owner executes a contract by engaging a consulting engineer to do the design work, preparing the tender documents, and generally also supervising the works during construction. After selecting a contractor, he would normally work under a direct contract for the employer. In this role, he is responsible for the execution of the works according to the contract conditions and specifications, and to instructions of the engineer who supervises the works.

In other contract forms, such as BOT (build, operate, transfer) contracts, the traditional role of the engineer has changed. The employer is often a group of private companies that are willing to finance a specific project on behalf of the government. Following a bidding procedure, it would order a contractor or joint venture of different contractors to realize the project, often on a turnkey basis, according to their own design which would normally be subcontracted to an engineer who might also supervise the construction works. It is quite obvious that there has been a change in the responsibilities of the engineer and particularly the key risk of the employer, that is, the responsibility for unexpected, unknown, and unforeseen ground conditions.

Traditionally, the engineer, working directly for the employer, would provide the contractor also with the results of the geotechnical investigations and would base his design on his own interpretation of these findings, which will also direct any contractor's own design work. Accordingly, it is the risk of the employer and his engineer if a substantial change of the geological conditions requires adaptations of the design or the work procedure. In these circumstances, it is the contractor's responsibility to notify the engineer of these changed conditions and, ultimately, the employer would have to reimburse the contractor for any additional work resulting from these changes. This principle should also apply if a contract was executed on a lump-sum basis with warranted dimensions and properties.

Where BOT contracts are concerned the employer's risk for the geotechnical conditions is passed on to the contracting consortium, which has to carry out all necessary investigations for the design and execution of the works and which employs therefore its own engineer.

In a majority of contracts, the necessary ground improvement works will be carried out by specialist contractors working as subcontractors. With both types of contracts, the geotechnical investigation and its competent interpretation by the engineer (either working on behalf of the employer or for the turnkey contractor) is essential for a successful execution of the foundation works. Regardless of who ultimately will bear the risk for the subsoil conditions, it is in the best interest of geotechnical engineers to help to make specialist foundation works such as ground improvement by deep vibratory compaction a technically and scientifically accepted method.

Contract conditions will also contain a project description including a time schedule to allow the specialist contractor to define the required plant and equipment. A bill of quantities with payment and legal conditions are the key elements of the contract which is generally complemented by technical specifications. The rules of the contract have to reflect the employer's right to have all works done according to the current state of the art.

The increasing importance of ground improvement in general, and the rising demand for the deep vibratory methods of vibro compaction and vibro replacement stone columns in particular, prompted engineering societies in many countries to work out standards, specifications, and notes of guidance to support good practice in this specialized field of construction industry (ICE, 1987; BRE, 2000; CFMS, 2005 or codes such as DIN EN 14731). In the United States, preference is given to case-by-case specifications that often accommodate the needs of the specific project in a better way. These and the guideline specifications generally reflect experts' knowledge in the technical field as well as in contractual affairs and serve for both design-and-build and for measured contracts.

The previous chapters have highlighted the importance of a comprehensive soil report which is prerequisite for the ground improvement design. It should contain all information on the soil which is generally necessary for bearing capacity and settlement calculations under the loading conditions of the project. When vibro compaction of loose granular soils is envisaged, the soil investigation has to include, apart from general information on in-situ densities and water table, stratification, sieve analyses, and mineralogical composition of the sand layers. We have seen that unsuitable cohesive soils are not just bypassed with the vibro stone column method. Their strength and deformation characteristics are essential information for the design. Equally important are the contractual regulations to cope with nonconformance situations when assessing the effectiveness of the ground improvement method. Whenever the deviations from the performance requirements are not caused by the quality of the works, specialist contractor and engineer together have to work out a solution to remedy this situation which will have to rely on an accurate description and of soil properties and their meaningful interpretation.

On vibro compaction contracts, performance checks are normally carried out by comparison of pre- and post compaction in-situ density measurements, such as cone penetration test (CPT) or standard penetration test, with specified values. Depending on the size of the projects, in-situ testing should be carried out at a frequency between 400 and 900 m² per test, with a time lapse for the posttreatment test of 2 weeks for silica and at least 3 weeks for calcareous sand. Although the location of the post-compaction test is generally chosen at the *weakest* point of the vibro compaction probe grid, other criteria for the location of the verification testing, particularly for large grid spacing in calcareous soils, were found to be more appropriate. To cope with the relatively sharp density attenuation in these soils with increasing distance from the compaction center, the weighted average better reflects the true compaction result. This can be obtained as shown in Section 3.6.7 or, somewhat less laborious, by calculating the average $q_{c(wa)}$ out of only two CPT measurements carried out at the *third point*, being representative for 40% of the grid area, and at the *mid point*, representing 60% of the area, of the distance between two compaction probes (see Equation 7.1 and Figure 7.1).

$$q_{c(wa)} = 0.4 \cdot q_{c(1/3)} + 0.6 \cdot q_{c(1/2)} \tag{7.1}$$

In addition to this approach, it is recommended to smoothen the q_c values by averaging the measurements over a depth increment of 0.3–0.5 m. The curve so developed can then be compared with performance criteria of the contract. A failed test would generally trigger verification procedures which ultimately could require remedial measures. Verification should be done generally by two additional CPTs which, when one of these fails again, would require remedial action, which could be recompaction or, in the case of adverse soil conditions, a detailed geotechnical analysis necessitating other ground improvement measures (vibro stone columns, compaction grouting, etc.).

The technical specifications should also contain detailed requirements for the compaction probe records, information on adjacent structures needing protection (underground services, seawalls, quays, and other buildings), details for the posttreatment surface grading and compaction, and a survey of the site levels before and after compaction, which is an excellent indicator of the effectiveness of the vibro compaction method.

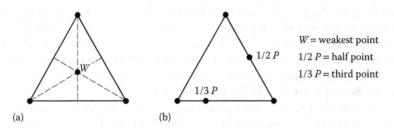

Figure 7.1 Location of cone penetration test in triangular grids: (a) silica sand and (b) calcareous sand.

Where vibro replacement stone columns are concerned, the specifications should not give too many details unless trial stone columns have been constructed beforehand, allowing the specialist contractor sufficient latitude in choosing the appropriate equipment and in the detailed construction procedure (Barksdale and Bachus, 1983). This philosophy is particularly valid when looking at the range of depth vibrators in use today. End result specifications therefore appear to be preferable restricting the key requirements to a certain design bearing pressure and providing limits for the total and differential settlement.

The construction process will, in the majority of projects, involve the dry bottom feed method since the use of jetting water necessary for wet stone columns requires special considerations in their construction (sludge handling and disposal of process water), and this is only rarely possible due to environmental restrictions. In the case of soft ground extending to ground surface, the owner and the engineer need to address the requirement of a granular working platform which is necessary for safe operation of the heavy equipment and the efficient execution of the work.

An effective data acquisition system recording all relevant process parameters, in particular the actual amount of stone backfill used, is indispensable for all vibro stone column projects. A realistic measurement of the stone volume over depth, currently state of the art with the bottom feed method, enables determination of the stone column diameter. It must be remembered that the column diameter and the friction angle of the stone column material are the key parameters determining their behavior under load. Stone consumption is therefore an essential element of quality control procedure and a key assumption made in the design. Measurement and control of the stone quantity is particularly important in very soft soils in which an overtreatment is to be avoided which otherwise may lead to an unwanted reduction of local strength of the surrounding soil by remolding. Similarly important are measures to avoid excessive heave, particularly in the final stage of column construction. Pre-boring, which can also become necessary to penetrate obstructive layers, is an effective measure in these instances.

Load testing will not be required when the performance of stone columns can be assessed with sufficient confidence from similar projects with comparable ground and loading conditions. In all other cases, and particularly for large contracts, load test on single columns, or on groups of columns—known as zone tests—is recommended and will help to verify bearing capacity and settlement performance requirements. Procedures for testing and instrumentation need to be specified together with the observation of the pore water pressure development in the ground to allow time to elapse before commencing the load test procedures.

Measurement of vibro compaction is generally done by the volume in cubic meter of sand to be compacted to a specified density. Pre- and posttesting at specified frequencies can also be included in this item. The compaction work includes rough site leveling. It may or may not include surface compaction of the upper 0.5 m. The engineer may specify a depth vibrator with a certain power rating, although this selection is better left to the specialist contractor and will be reflected in his proposal by the probe pattern, method statement, and execution time necessary for the volume of works. For preliminary assessment of the time to compact sand from loose to medium density, an average production rate of about 7500 m^3 for a 10-h shift of a single depth vibrator is reasonable. Production and unit price depend, of course, on many parameters and conditions described in Chapter 3.

For vibro stone columns the proposed unit for measurement is lin.m of stone column with specified diameter, length, and quality of gravel backfill, which the contractor has to identify with appropriate data acquisition, calibrated before starting the work and at suitable intervals during the contract. The specifications will indicate which method of execution (wet, dry top or bottom feed system) should be used. It is advisable to specify a minimum power rating for the depth vibrator to be used to ensure proper densification of the stone backfill. The decision to pre-bore through hard layers should be left with the specialist contractor, provided their strength can be described sufficiently accurately. For normal conditions, an average production rate of 300 lin.m of 0.6 m diameter dry bottom feed stone columns can be assumed for a 10-h shift and one machine operating.

The main item for the execution of stone columns is occasionally split in an item for the column construction in lin.m and a second item for the delivery and placement of the stone backfill in cubic meter or per ton. To underline the importance of the stone column diameter, the authors do recommend always specifying the stone columns together with a minimum diameter according to the design which has to be verified during execution of the works. Price of the backfill stone can be decisive for the cost of a vibro replacement project. It is therefore important to find suitable material at economic prices. The use of sand or recycled material is often a worthwhile compromise, even if the stone column design needs amending for reduced shear properties.

Vibro concrete columns represent a variation of small diameter in-situ concrete piles and should, from a contractual point of view, be considered accordingly. The establishment and on-site control of the load carrying capacity and settlement performance should generally be by standard load testing, only in exceptional cases and with sufficient experience should dynamic load testing be performed. Vibro concrete column foundations should only be carried out by experienced contactors using efficient data acquisition systems closely controlling especially verticality and diameter of the column.

In all cases in which the works are being executed according to the specialist subcontractor's design, the employer or turnkey contractor will base his decision for the contract award on price, time of execution, and a warranted design. Resolution of disputes on the extent of the treatment (probe spacing and depth, column diameter, depth, and distance) can only be on the basis of a comprehensive geotechnical report.

References

Aboshi, H. et al. (1979) The composer: A method to improve characteristics of soft clays by inclusion of large diameter sand columns. In *Colloque International sur le Renforcement des Sols*. ENPC-LCPC, Paris, France.

Achmus, M. et al. (2007) Untersuchung zu Bauwerks- und Bodenerschütterungen infolge Tiefenrüttlung. In *3. Hans Lorenz Symposium*, Grundbauinstitut H. 41, TU Berlin, Berlin, Germany.

Achmus, M. et al. (2010) Building vibrations due to deep vibro processes. In *7th Conference on Ground Improvement Techniques*. Seoul, South Korea.

Adalier, K. et al. (2003) Stone columns as liquefaction counter measure in nonplastic silty soils. *Soil Dynamics and Earthquake Engineering*, 23(7).

Ahrens, W. (1941) Die Bodenverdichtung bei der Gründung für die Kongresshalle in Nürnberg. *Die Bauindustrie*, 9(35).

Arnold, M. et al. (2008) Comparison of vibrocompaction methods by numerical simulations. In Karstanen, M. et al. (eds), *Geotechnics of Soft Soil: Focus on Ground Improvement. 2nd International Workshop on Geotechnics of Soft Soils*. University of Strathclyde, Glasgow, UK.

Avalle, D. (2007) *Trials and Validation of Deep Compaction Using the "Square" Impact Roller*. Advances in Earthwork, Australian Geomechanics Society, Sydney, Australia.

Baez, J.I. (1995) A design model for the reduction of soil liquefaction by vibro stone columns. PhD Thesis, University of Southern California, Los Angeles, CA.

Baez, J.I. and Martin, G.R. (1993) Advances in the design of vibro systems for the improvement of liquefaction resistance. In *Proceedings of the Symposium on Ground Improvement*. Vancouver Geotechnical Society, Vancouver, Canada.

Baez, J.I. et al. (1998) Liquefaction mitigation of silty dam foundation using vibro stone columns and drainage wick drains: A test section case history at Salmon Lake Dam. In *Proceedings of Association of State Dam Safety Officials Annual Conference*, Las Vegas, NV.

Baez, J.I. et al. (2000) Comparison of SPT-CPT liquefaction evaluations and CPT interpretations, innovations and applications in geotechnical site characterization. In *Proceedings of Sessions of Geo-Denver 2000*. ASCE GSP 97, Denver, CO.

Balaam, N.P. and Booker, J.R. (1981) Analysis of rigid rafts supported by granular piles. *International Journal for Numerical and Analytical Methods in Geomechanics*, 5, pp. 379–403.

Balaam, N.P. and Booker, J.R. (1985) Effect of stone column yield on settlement of rigid foundations in stabilized clay. *International Journal for Numerical and Analytical Methods in Geomechanics*, 9(4), 331–351.

Balaam, N.P. et al. (1977) *Settlement Analysis of Soft Clays Reinforced with Granular Piles*. School of Engineering, The University of Sydney, Sydney, Australia.

Balaam, N.P. and Poulos, H.G. (1983) The behaviour of foundations supported by clay stabilised by stone columns. In *Proceedings of the VIIIth ICSMFE.* Helsinki, Finland.

Barksdale, R.D. and Bachus, R.C. (1983) *Design and Construction of Stone Columns*. FHWA/RD-83/026, US Department of Transportation, Georgia Institute of Technology, Atlanta, GA.

Barksdale, R.D. and Bachus, R.C. (1984) Vertical and lateral behaviour of model stone columns. In *Proceedings of the International Conference on In-Situ Soil and Rock Reinforcement*. Presses de l'ecole nationale des ponts et chausses: Paris, France.

Barron, R.A. (1948) Consolidation of fine-grained soils by drain wells. *Transactions of ASCE*, 113(2346), 718–742.

Bauldry, B.D. et al. (2008) Monitoring for successful site improvement. In *Geotechnical Earthquake Engineering and Soil Dynamics IV*. ASCE GSP 181, Reston, VA.

Bell, A.L. (1915) The lateral pressure and resistance of clay and the supporting power of clay foundations. *Proceedings of the Institution of Civil Engineers*, 199, 233–272.

Bellotti, R. and Jamiolkowski, M. (1991) Evaluation of CPT and DMT in crushable and silty sands. Third interim report ENEL CRIS, Milan, Italy.

Bergado, D.T. et al. (1994) *Improvement Techniques of Soft Ground in Subsiding and Lowland Environment*. A.A. Balkema, Rotterdam, the Netherlands.

Betterground (2014) Quality Control and Operator Guidance Systems for Stone Columns.

Bohn, C. et al. (2016) Bestimmung von Mantelreibung und Spitzendruck von Betonstopfsäulen von gemessenen Lastsetzungskurven. Beitrag zum *23. Darmstädter Geotechnik- Kolloquium*. Mitteilungen des Instituts und der Versuchsanstalt für Geotechnik der Technischen Universität Darmstadt.

Boley, C. (2007) Filter stability of stone columns. Universität der Bundeswehr München. Commissioned by Keller (unpublished).

Borchert, K.-M. et al. (2005) Berechnung und Ausführung einer Rüttelstopfverdichtung in weichem Geschiebemergel. Beitrag zum *4. Geotechniktag München*. Schriftenreihe des Lehrstuhls für Bodenmechanik und Felsmechanik der TU München, 37.

Borchert, K.-M. et al. (2005) Betonsäulen als pfahlartige Tragglieder–Herstellungsverfahren, Qualitätssicherung, Tragverhalten und Anwendungsbeispiele. *Pfahl-Symposium 2005*. Mitteilungen des Instituts für Grundbau und Bodenmechanik der TU Braunschweig, 80.

Brauns, J. (1978) Die Anfangstraglast von Schottersäulen im bindigen Untergrund. *Die Bautechnik*, 8.

Brauns, J. (1980) Untergrundverbesserung mittels Sandpfählen oder Schottersäulen. *Tiefbau, Ingenieurbau, Straßenbau*, 22(8), 678–683.

BRE Building Research Establishment. (2000) *Specifying Vibro Stone Columns*. CRC Press, Watford, England.

Breitsprecher, G. et al. (2009) Flughafen Berlin Brandenburg International-Dimensionierung und Ausführung einer Baugrundverbesserung für Bauwerke und Verkehrsflächen der landseitigen Anbindung. In *5. Hans Lorenz Symposium*, TU Berlin, Berlin, Germany.

Bremer, K. and Hofmann, O.E. (1976) Tiefenverdichtung von gleichförmigen Sanden mit Tauchrüttlern. *Tiefbau, Ingenieurbau, Strassenbau*, Juli.

Breth, H. (1973) Die Verflüssigung wassergesättigter Sande die Möglichkeit, ihrer zu begegnen; ein Beitrag zur Reaktorsicherheit. In *Festschrift zum 60*, Geburtstag von Professor Börnke, Essen, Germany.

Brown, R.E. (1977) Vibroflotation compaction of cohesionless soils. *Journal of Geotechnical Engineering*, ASCE, 103(GT12), 1437–1451.

BSI British Standards Institution. (2008) *PAS 2050:2008-Specifications for the Assessment of the Lifecycle Greenhouse Gas Emissions of Goods and Services.*

Bureau of Reclamation. (July 20, 1948) *Vibroflotation Experiments at Enders Dam.* Bureau of Reclamation, Denver, CO.

Casagrande, A. (1936) Characteristics of cohesionless soils affecting the stability of slopes and earthfills. *Journal of the Boston Society of Civil Engineers*, 23.

CFMS Comite Francais de Mecanique des Sols (2005) Recommandations sur la conception, le calcul, l'exécution et le controle des colonnes ballastees sous batiments et ouvrages sensibles au tassement. *Revue Francaise de Geotechnique*, 111.

Charlie, W.A. et al. (1992) Time-dependent cone penetration resistance due to blasting. *Journal of Geotechnical Engineering*, ASCE, 118(8), 1200–1215.

Committee on Placement and Improvement of Soils (1978) *Soil Improvement: History, Capabilities and Outlook.* GED of the ASCE, New York.

Cudmani, R.O. (2001) Statische, alternierende und dynamische Penetration in nicht bindigen Böden. Dissertation, Karlsruhe University, Karlsruhe, Germany.

CUR (1996) *Building on Soft Soils.* Centre for Civil Engineering Research and Codes. A.A. Balkema, Rotterdam, the Netherlands.

D'Appolonia, E. (1954) *Loose Sands: Their Compaction by Vibroflotation.* ASTM No. 156 Special Technical Publication.

Das, B.M. (1993) *Principles of Soil Dynamic.* PWS-Kent Publishing Company, Boston, MA.

Debats, J.M. et al. (2003) Soft soil improvement due to vibro compacted columns installations. In Vermeer, P.A., Schweiger, H.F., Karstunen, M., and Cudny, M. (eds), *International Workshop on Geotechnics of Soft Soils: Theory and Practice.* VGE, Essen, Germany.

Degebo. (1940) Versuchsbericht (unpublished).

Degebo. (1942) Gutachten über die Tragfähigkeit des nach dem Kellerverfahren verfestigten Untergrundes in Rolingheten/Heroen (unpublished).

Degen, W. (1997a) 56 m Deep vibro-compaction at German lignite mining area. In *Proceedings of the 3rd International Conference on Ground Improvement Geosystems*, Heron Quays, London, UK.

Degen, W. (1997b) Vibroflotation Ground Improvement (unpublished).

Degen, W.S. (2014) Advances in equipment and quality control for offshore and onshore stone columns and vibro compaction. In *18th Annual National Conference on Geotechnical Engineering*, Jakarta, Indonesia.

DGGT. (in press) Empfehlungen des Arbeitskreises 2.8 (Stabilisierungssäulen) der Deutschen Gesellschaft für Geotechnik e.V. German Geotechnical Society, Essen, Germany.

Dücker, F.-J. (1957) Über Erddruckmessungen an einem Brückenwiderlager, *Die Bautechnik*, 11.

Dücker, F.-J. (1968) Bodenverdichtung und Gründungen unter Verwendung von Kellerschen Grossrüttlern. In *5. Internationale Hafenkongress*, Antwerpen, Belgium.

EFFC-DFI. (2013) *Carbon Calculator Methodological & User Guide*. Benoit Lemaignan & Jean Yves Wilmotte. http://www.carbone4.com/.

Egan, D. and Slocombe, B. (2010) Demonstrating the environmental benefits of ground improvement. *Ground Improvement. Proceedings of the Institution of Civil Engineers*, 163(1), 63–69.

Engelhardt, K. and Golding, H.C. (1973) Field testing to evaluate stone column performance in a seismic area. *Geotechnique*, 25(1), 61–69.

European Committee for Standardisation. (2005) Execution of special geotechnical works—Ground treatment by deep vibration, EN 14731:2005E CEN, Brussels.

Fellin, W. (2000) Rütteldruckverdichtung als plastodynamisches Problem. *Advances in Geotechnical Engineering and Tunnelling*, 3, 344pp (February).

Finn, W. D. L. et al. (1971) Sand liquefaction in triaxial and simple shear tests. *Journal of Soil Mechanics and Foundation Division*, ASCE, 97(SM4), 639–659.

Floroui, L. and Schweiger, H. (2015) Parametric study of the seismic ground response of a linear visco-elastic soil layer improved by stone columns or pile-like elements, *Geotechnik*, 28(Heft 4), 304–315, Ernst & Sohn, Berlin, Germany.

Forrest, M. et al. (2004) Stone column construction stabilizes liquefiable foundation at Lopez dam. In *Proceedings of Dam Safety. ASDSO's 21st Annual Conference*, Phoenix, AZ.

Frankipile Australia. (November 1974) Grainterminal Kwinana, Western Australia. Foundation Works. Contracting and Construction Engineer. http://www.franki.com.au.

Gibson, R.E. and Anderson, W.F. (1961) In-situ measurement of soil properties with the pressuremeter. *Civil Engineering and Public Work Review*, 56(658), 615–618.

Goughnour, R.R. (1983) Settlement of vertically loaded stone columns in soft ground. In *Proceedings of the 8th ECSMFE*, vol. 1, Helsinki, Finland.

Goughnour, R.R. and Bayuk, A.A. (1979a) A field study of long-term settlements of loads supported by stone columns in soft ground. In *Colloque International Renforcement des Sols*, Paris, France.

Goughnour, R.R. and Bayuk, A.A. (1979b) Analysis of stone column–soil matrix interaction under vertical load. In *Colloque International sur le Renforcement des Sols*, Paris, France.

Greenwood, D.A. (1970) *Mechanical Improvement of Soils below Ground Surface*. Ground Engineering. The Institution of Civil Engineers, London, UK.

Greenwood, D.A. (1976) *Ground Treatment by Deep Compaction*. Discussion. The Institution of Civil Engineers, London, UK, 123p.

Greenwood, D.A. (1991) Load tests on stone columns. In Esrig, M.I. and Bacchus, R.C. (eds), *Deep Foundation Improvements: Design, Construction and Testing*. ASTM STP 1089, ASTM, Philadelphia, PA.

Greenwood, D.A. and Kirsch, K. (1984) *Specialist Ground Treatment by Vibratory and Dynamic Methods. Piling and Ground Treatment*. The Institution of Civil Engineers, Thomas Telford, London, UK.

Gruber, F.J. (1994) Verhalten einer Rüttelstopfverdichtung unter einem Straßendamm. Dissertation, TU Graz, Styria, Austria.

Herle, I. et al. (2007) Einfluß von Druck und Lagerungsdichte auf den Reibungswinkel des Schotters in Rüttelstopfsäulen. *Pfahl Symposium*. Institut für Grundbau und Bodenmechanik. TU Braunschweig, Braunschweig, Germany, p. 84.

Hoffmann, R. and Muhs, H. (1944) Die mechanische Verfestigung sandigen und kiesigen Baugrundes. *Die Bautechnik*, 22(Heft 33/36), 149–155.

Hu, W. (1995) *Physical Modelling of Group Behaviour of Stone Column Foundations*. Dissertation. University of Glasgow, Glasgow, UK.

Hughes, J.M. and Withers, N.J. (1974) Reinforcing of soft cohesive soils with stone columns. *Ground Engineering*, 7(3).

Hughes, J.M. et al. (1975) A field trial of the reinforcing effect of a stone column in soil. *Geotechnique*, 25(1), 31–44.

Hussin, J.D. (2006) Methods of soft ground improvement. *The Foundation Engineering Handbook*. Taylor & Francis Group, Boca Raton, FL.

ICE (Institution of Civil Engineers). (1987) *Specification for Ground Treatment*. Thomas Telford, London, UK.

Indian Standard (IS) 15284. (2003). Indian standard code of practice for design and construction for ground improvement guidelines.

IREX. (2013) Institut pour la recherche appliquée et l'expérimentation en génie civil. *Recommandations for the design, construction and control of rigid inclusion ground improvements*. Presses des Ponts, Paris, France.

Jebe, W. and Bartels, K. (1983) Entwicklung der Tiefenverdichtungsverfahren mit Tiefenrüttlern von 1976–1982. In *VIIIth ECSMFE*, Helsinki, Finland.

Johann Keller GmbH. (1935) *Der Rütteldruck. Die neue Technik des Erd- und Betonbaus*. Company Publication.

Johann Keller GmbH. (1936) *Der Rütteldruck. Die neue Technik des Erd- und Betonbaus in Anwendung auf die Verdichtung des Baugrundes beim Kongressbau Nürnberg*. Company Publication.

Johann Keller GmbH. (1938) *Bodenverfestigung nach dem Rütteldruckverfahren (DRP und Auslandspatente)*. Company Publication.

Johann Keller GmbH. (1958) *Verdichtungsversuche für den Bau des Hochdammes bei Assuan*. Unpublished report for the Sadd el Aali Authority, Cairo, Egypt.

Kempfert, H.-G. and Stadel, M. (1995) Zum Tragverhalten geokunststoffbewehrter Erdbauwerke über pfahlähnlichen Traggliedern. Geotechnik Sonderheft zur 4. *Informations- und Vortragsveranstaltung über Kunststoffe in der Geotechnik*. Verlag DGGT, München, pp. 146–152.

Killeen, M. (2012) Numerical modelling of small groups of stone columns. PhD Thesis, National University of Ireland, Galway, Ireland.

Killeen, M. and McCabe, B.A. (2009) A numerical study of factors governing the performance of stone columns supporting rigid footings on soft clay. In *7th European Conference on Numerical Methods in Geotechnical Engineering*, Trondheim, Norway. Taylor & Francis Group, Boca Raton, FL.

Kirsch, K. (1979) Erfahrungen mit der Baugrundverbesserung durch Tiefenrüttler. *Geotechnik*, 1.

Kirsch, K. (1985) Application of deep vibratory compaction in harbour construction. In *Egyptian-German Seminar on Foundation Problems in Egypt and Northern Germany*, Cairo, Egypt.

Kirsch, K. (1985) Over 50 years of deep vibratory compaction: Milestones of German geotechnique. *Geotechnik*, Special Issue, Deutsche Gesellschaft für Erd- und Grundbau, Essen, Germany.

Kirsch, K. (1993) Die Baugrundverbesserung mit Tiefenruettlern. In Englert, K. and Stocker, M. (eds), *40 Jahre Spezialtiefbau 1953–1993: Technische und Rechtliche Entwicklungen*. Werner Verlag, Düsseldorf, Germany.

Kirsch, F. (2004) Experimentelle und numerische Untersuchungen zum Tragverhalten von Rüttelstopfsäulengruppen. Dissertation, Technische Universität Carolo-Wilhelmina zu Braunschweig, Braunschweig, Germany.

Kirsch, F. (2006) Vibro stone column installation and its effect on ground improvement. In Triantafyllidis, Th. (ed.), *Numerical Modelling of Construction Processes in Geotechnical Engineering for Urban Environment*. Taylor & Francis Group, London, UK.

Kirsch, F. (2008) Evaluation of ground improvement by groups of vibro stone columns using field measurements and numerical analysis. In Karstunen, M. et al. (eds), *Geotechnics of Soft Soils-Focus on Ground Improvement, Proceedings of the 2nd International Workshop on Geotechnics of Soft Soils*, Glasgow, Scotland. CRC Press, Boca Raton, FL.

Kirsch, F. und Borchert, K.-M. (2006) Probebelastungen zum Nachweis der Baugrundverbesserungswirkung. In *21. Chr. Veder Kolloquium: Neue Entwicklungen der Bodenverbesserung*, TU Graz, Graz, Austria.

Kirsch, F. and Sondermann, W. (2001) Ground improvement and its numerical analysis. In *Proceedings of the XVth ICSMFE*, Istanbul, Turkey. A.A. Balkema, Rotterdam, the Netherlands.

Kirsch, F. and Sondermann, W. (2003) Field measurements and numerical analysis of the stress distribution below stone column supported embankments and their stability. In Vermeer, P.A. et al. (eds), *Geotechnics of Soft Soils: Theory and Practice*. VGE, Essen, Germany.

Kirsch, F. et al. (2004) Berechnung von Baugrundverbesserungen nach dem Rüttelstopfverfahren. In *Vorträge der Baugrundtagung 2004 in Leipzig*. Hrsg. DGGT, VGE, Essen, Germany.

Kolymbas, D. and Fellin, W. (2000) *Compaction of Soils, Granulates and Powders: Advances in Geotechnical Engineering and Tunnelling*. A.A. Balkema, Rotterdam, the Netherlands.

Kutzner, C. (1996) *Grouting of Rock and Soil*. A.A. Balkema, Rotterdam, the Netherlands.

Lackner, E. (1966) Schwierige Gründungen in Verbindung mit Bodenverbesserungen. *Der Bauingenieur*, 41(9).

Lawton, G.M. et al. (2004) First use of stone columns in California under state regulatory jurisdiction—The seismic remediation design of Lopez Dam. In *Proceedings of Dam Safety. ASDSO's 21st Annual Conference*, Phoenix, AZ.

Lee, J.S. and Pande, G.N. (1998) Analysis of stone column reinforced foundations. *International Journal Numerical Analytical Methods Geomechanics*, 22.

Lee, J.S. et al. (1999) Elasto-plastic analysis of composite material using a macro level yield function. In Pande, G.N., Pietruszczak, S., and Schweiger, H.F. (eds), *Numerical Models in Geomechanics—NUMOG VII*. A.A. Balkema, Rotterdam, the Netherlands.

Liquefypro. (2008) *Liquefaction and Settlement Analysis*. Software Manual.

Loos, W. (1936) Comparative studies of the effectiveness of different methods for compaction cohesionless soils. In *Proceedings of the 1st ICSMFE*, vol. 3, Cambridge, MA, pp. 174–182.

Lopez, R.A. and Shao, L. (2008) Use of the static and seismic deformation criteria for vibro replacement stone columns: A case history. In *ASCE GSP 172 Soil Improvement*.

Lunne, T. and Christophersen, H.P. (1983) Interpretation of cone penetrometer data for offshore sands. In *Proceedings of the Offshore Technology Conference*, Richardson, TX, Paper No. 4464.

Lunne, T. et al. (1997) *Cone Penetration Testing in Geotechnical Practice*. Blackie Academic & Professional, Glasgow, Scotland.

Luongo, V. (1992) *Predicting Depth of Improvement in Grouting, Soil Improvement and Geosynthetics*, ASCE Geotechnical Special Publication No. 30, New Orleans, LA.

Massarsch, K.R. (1991) *Deep Soil Compaction Using Vibratory Probes in Deep Foundation Improvement*, STP 1089, ASTM.

Massarsch, K.R. (May 19, 1994) Design aspects of deep vibratory compaction. In *Proceedings of Seminar on Ground Improvement Methods*, Geotechnical Division, Hong Kong Institution of Engineers, Hong Kong, China, 61–74.

Massarsch, R.K. and Fellenius, B.H. (2005) Deep vibratory compaction of granular soils. In Indranatna, B. and Jian, C. (eds), *Ground Improvement—Case Histories*. Elsevier Publishers, pp. 633–658 (Chapter 19).

Maurer, C. (2008a) Influence of vibrations on adjacent buildings and soil as a result of deep vibro processes. Internal Report. Keller Holding GmbH.

Maurer, C. (2008b) Sound measurements during vibro replacement in Germany and England. Internal Report. Keller Holding GmbH.

Mitchell, J.K. (1984) Practical problems from surprising soil behaviour. 12th Terzaghi Lecture. *Journal of Geotechnical Engineering*, ASCE, 112(3).

Mitchell, J.K. (2013) *GeoTechTools—An Interactive Web-based Information and Guidance System for Ground Improvement Technologies*. Presented at the National Center for Researches in Earthquake Engineering, Taipei, Taiwan.

Mitchell, J.K. and Katti, R.K. (1981) Soil improvement: State of the art report. In *Xth ICSMFE*, Stockholm, Sweden.

Mitchell, J.K. and Solymar, Z.V. (1984) Time-dependent strength gain in freshly deposited or densified sand. *Journal of Geotechnical Engineering*, ASCE, 110(11).

Moseley, M.P. and Priebe, H.J. (1993) Vibro techniques. In Moseley, M.P. (ed), *Ground Improvement*. Blackie Academic & Professional, Glasgow, Scotland.

Muhs, H. (1949) Arbeiten der Degebo in den Jahren 1938–1948. *Bautechnik-Archiv*, Heft 3, 20–40.

Muhs, H. (1969) 1928–1968 40 Jahre Degebo. *Mitteilungen der Deutschen Gesellschaft für Bodenmechanik (Degebo) an der Technischen Universität Berlin*, Heft 23, Berlin, Germany.

Muir Wood, D. et al. (2000) Group effects in stone column foundations: Model tests. *Geotechnique*, 50(6), 689–698.

Nahrgang, E. (1976) Untersuchung des Tragverhaltens von eingerüttelten Schottersäulen an Hand von Modellversuchen. *Baumaschine + Bautechnik*, 23(Heft 8), 391–404.

NCEER (1997) In Youd, T.L. and Idriss, I.M. (eds), *Proceedings of the NCEER Workshop on Evaluation of Liquefaction Resistance*. Technical Report NCEER-97-0022.

Nendza, M. (2007) Untersuchungen zu den Mechanismen der Dynamischen Bodenverdichtung bei Anwendung des Rütteldruckverfahrens. Dissertation, Technische Universität Carolo—Wilhelmina zu Braunschweig, Braunschweig, Germany.

O'Hara, V. (2003) Vibration and noise levels on ground improvement sites. Guidance Notes for Engineers. Keller Ground Engineering (unpublished).

O'Hara, V. and Davison, A. (2004) Noise propagation from driven piling. Noise control engineering. Inter-Noise 2004. Prague, Czech Republic.

Österreichische Gesellschaft für Geomechanik. (2013) Empfehlungen für die Ausschreibung von Tielenrüttelverfahren (Rüttelstopf-/Rütteldruckverdichtungen). salzburg@oegg.at.

Parsons-Brinkerhoff et al. (1980) Jourdan Road Terminal Test Embankment. Report Prepared for Board of Commissioners of the Port of New Orleans, New Orleans, LA.

Plannerer, A. (1965) Das Rütteldruckverfahren. Seine Weiterentwicklung und Anwendung für Gründungsaufgaben. Report of the Institute for Soil Mechanics and Foundation Engineering, Technical University Vienna, Wien, Austria.

Priebe, H.J. (1976) Abschätzung des Setzungsverhaltens eines durch Stopfverdichtung verbesserten Baugrundes. Die Bautechnik, 53(8).

Priebe, H.J. (1987) Abschätzung des Scherwiderstandes eines durch Stopfverdichtung verbesserten Baugrundes. Die Bautechnik, 55(8).

Priebe, H.J. (1988) Zur Abschätzung des Setzungsverhaltens eines durch Stopfverdichtung verbesserten Baugrundes. Die Bautechnik, 65(1).

Priebe, H.J. (1995) Die Bemessung von Rüttelstopfverdichtungen. Die Bautechnik, 72(3).

Priebe, H.J. (2003) Zur Bemessung von Rüttelstopfverdichtungen—Anwendung des Verfahrens bei extrem weichen Böden, bei schwimmenden Gründungen und beim Nachweis der Sicherheit gegen Gelände—oder Böschungsbruch. Die Bautechnik, 80(6), 380–384.

Priebe, H.J. (2014) Die Setzung von Fundamenten auf unterschiedlichen Gründungen mit eingeschränkter Lastausbreitung. Bautechnik 91(Heft 6).

Raithel, M. (1999) Zum Trag- und Verformungsverhalten von geokunststoffummantelten Sandsäulen, vol. 6. Schriftenreihe Geotechnik, Universität Kassel, Kassel, Germany.

Raithel, M. and Kempfert, H.-G. (1999) Bemessung von geokunststoffummantelten Sandsäulen. Die Bautechnik, 76(11), 983–991.

Raju, V.R. et al. (2004) Vibro replacement for the construction of a 15 m high highway embankment over a mining pond. In Malaysian Geotechnical Conference, Kuala Lumpur, Malaysia.

Raju, V.R. and Hoffmann, G. (1996) Treatment of tin mine tailings in Kuala Lumpur using vibro replacement. In Proceedings of the 13th Southeast Asian Geotechnical Conference. Kuala Lumpur, Malaysia.

Raju, V.R. and Wegner, R. (1998) Ground improvement using vibro techniques: Case histories from South East Asia. In Ground Improvement Conference, Singapore.

Rappert, C. (1952) Die Entwicklung von Großrüttlern und ihre Einsatzmöglichkeiten im Talsperrenbau. Die Wasserwirtschaft, 4.

Rausche, F. et al. (1985) Dynamic determination of pile capacity. Journal of Geotechnical Engineering Division, ASCE, 111(9), 367–383.

Robertson, P.K. (1990) Soil classification using CPT. *Canadian Geotechnical Journal*, 27(1), 151–158.

Robertson, P.K. and Campanella, R.G. (1983) Interpretation of cone penetration tests. *Canadian Geotechnical Journal*, 20(4), 718–733.

Robertson, P.K. et al. (1983) SPT-CPT correlations. *Journal of Geotechnical Engineering*, 109(11), 1449–1459.

Rodger, A.A. (1979) Vibrocompaction of cohesionless soils, Internal Report, R.7/79. Cementation Research Limited, Croydon, UK.

Rodger, A.A. and Littlejohn, G.S. (1980) A study of vibratory driving in granular soils. *Géotechnique*, 30(3), 269–293.

Saito, A. et al. (1987) A countermeasure for sand liquefaction, "gravel drains method." Nippon Kokan Technical Report Overseas No. 51.

Savidis, S. (2007) *Grundbau-Dynamik*. FG Grundbau und Bodenmechanik-Degebo, Technische Universität Berlin, Berlin, Germany.

Scheidig, A. (1940) Speichergründung auf Rüttelfusspfählen. *Die Bautechnik*, 25.

Schmertmann, J.H. (1970) Static cone to compute static settlement over sand. *Journal of the SMFD*, ASCE, 96(3), 1011–1043.

Schmertmann, J.H. (1978) Guidelines for cone penetration test, performance and design. Report FHWA-TS-78-209, 145. US Federal Highway Administration, Washington, DC.

Schmertmann, J.H. (1991) The mechanical aging of soils. *Journal of Geotechnical Engineering*, ASCE, 117(12).

Schneider, H. (1938) Das Rütteldruckverfahren und seine Anwendung im Erd- und Betonbau. *Beton und Eisen*, Jahrgang, 37(Heft 1).

Schultze, E. and Moussa, A. (1961) Factors affecting the compressibility of sand. In *Proceedings of the 5th ICSMFE*, vol. 1, Paris, France.

Schweiger, H.F. (1989) Finite element analysis of stone column reinforced foundations. Dissertation, University of Swansea. Mitteilungen des Inst. für Bodenmechanik, Felsmechanik und Grundbau TU Graz. Heft 8.

Schweiger, H.F. (1990) Finite Element Berechnung von Rüttelstopfverdichtungen. In *5. Chr. Veder Kolloquium*. TU Graz, Styria, Austria.

Seed, H.B. and Booker, J.R. (1976) Stabilization of potentially liquefiable sand deposits using gravel drain systems. Report No. EERC 76-10. University of California, Berkeley, CA.

Seed, H.B. and Booker, J.R. (July, 1977) Stabilization of potentially liquefiable ground deposits using gravel drains. *Journal of Geotechnical Engineering Division*, ASCE, 103(GT 7).

Seed, H.B. and Idriss, I.M. (1971) Simplified procedure for evaluating soil liquefaction potential. *Journal of the SMFD*, ASCE, 97(SM9), 1249–1274.

Seed, H.B. and Idriss, I.M. (1982) *Ground Motions and Soil Liquefaction during Earthquakes*. Earthquake Engineering Research Institute Monograph.

Seed, H.B. et al. (1985) The influence of SPT procedures in soil liquefaction resistance evaluations. *Journal of Geotechnical Engineering*, ASCE, 111(12), 1425–1445.

Sehn, L.S. (2003) Shear strength of vibro replacement stone and how it affects deformation. In *Hayward Baker Engineering Conference*.

Shake (2000) A computer program for the 1-D analysis of geotechnical earthquake engineering problems.

Sidak, N. (2000) Soil improvement by vibro compaction in sandy gravel for ground water reduction. Compaction of soils, granulates and powders.

In Kolymbas, D. and Fellin, W. (eds), *Advances in Geotechnical Engineering and Tunnelling*. A.A. Balkema, Rotterdam, the Netherlands.

Silver, M.L. and Seed, H.B. (1971) Volume changes in sands during cyclic loading. *Journal of Soil Mechanics Foundation Division*, ASCE, 97(9).

Slocombe, B.C. and Smith, P. (2008) The impact of vibro compaction adjacent to large vertical maritime retaining structures. In *Proceeding of the 33rd Annual and 11th International Conference on Deep Foundations*, New York.

Smoltczyk, H.-U. (1966) Unterschätzen wir die Festigkeit des Sandes? *Wasser und Boden*. 9.

Solymar, Z.V. et al. (1984) Earth foundation treatment at Jebba Dam site. *Journal of Geotechnical Engineering*, 110(10), 1415–1430.

Sondermann, W. and Jebe, W. (1996) *Methoden zur Baugrundverbesserung für den Neu- und Ausbau von Bahnstrecken auf Hochgeschwindigkeit*, Vorträge der Baugrundtagung. Deutschen Gesellschaft für Geotechnik, Essen, Germany, pp. 259–279.

Sondermann, W. and Kirsch, K. (2009) Baugrundverbesserung. In *Grundbautaschenbuch, 7. Auflage. Teil 2: Geotechnische Verfahren*. Hrsg.: Witt, K.J., Ernst & Sohn, Berlin, Germany.

Sondermann, W. and Wehr, W. (2004) Deep vibro techniques. In Moseley, M.P. and Kirsch, K. (eds), *Ground Improvement*, 2nd edition. Spon Press, London, UK.

Sondermann, W. and Wehr, W. (2013) Kombinierte Gründung für eine Eisenerzverarbeitungsanlage. In *9. Hans Lorenz Symposium am Grundbauinstitut der TU Berlin*, Berlin, Germany.

Soyez, B. (April 1987) Bemessung von Stopfverdichtungen. *BMT*, 170–185.

Stern, N. (2006) Review of the economic challenges of climate change. OCC Analytical Audit. Executive Summary. http://webarchive.nationalarchives.gov.uk/+/http://www.hm-treasury.gov.uk/sternreview_index.htm.

Steuermann, S. (July, 1939) A new soil compaction device. *Engineering News Record*, p. 20.

Tanimoto, K. (1973) Introduction to the sand compaction pile method as applied to stabilisation of soft foundation grounds. Commonwealth Scientific and Industrial Research Organisation, Clayton South, Victoria, Australia.

Tate, T.N. (1961) World's largest drydock. *Civil Engineering*, pp. 33–37.

Terzaghi, K. and Peck, R.B. (1961) *Die Bodenmechanik in der Baupraxis*. Springer-Verlag, Berlin, Germany.

Thorburn, S. (1975) Building structures supported by stabilized ground. *Geotechnique*, 25.

Thorburn, S. et al. (1968) Soil stabilization employing surface and depth vibrators. *The Structural Engineer*, 46(10).

Thurner, R. and Kirsch, F. (2014) Combination of bored piles and stone columns for the earthquake resistant design. In *Second European Conference on Earthquake Engineering and Seismology*, Istanbul, Turkey.

Tokimatsu, K. and Seed, H.B. (1987) Evaluation of settlements in sand due to earthquake shaking. *Journal of Geotechnical Engineering*, ASCE, 113, 861–878.

Topolnicki, M. (2004) In situ soil mixing. In Moseley, M.P. and Kirsch, K. (eds), *Ground Improvement*. Spon Press, London, UK.

Topolnicki, M. (1996) Case history of a geogrid-reinforced embankment supported on vibro concrete columns. In deGroot, M.B. et al. (eds) *Geosynthetics: Application, Design and Constuction*. Balkema, Rotterdam, the Netherlands, pp. 333–340.

University Karlsruhe. (2006) CPT calibration testing on calcareous sands. Commissioned by Keller Grundbau. Unpublished.

van Impe, W.F. (2001) About the effectiveness of stone columns. In *Proceedings of the 15th. ICSMGE*, vol. 4, Istanbul, Turkey.

van Impe, W.F. and Debeer, E. (1983) Improvement of settlement behaviour of soft layers by means of stone columns. In *Proceedings of the 8th ECSMFE*, vol. 1, Helsinki, Finland.

van Impe, W.F. and Madhav, M.R. (1992) Analysis and settlement of dilating stone column reinforced soil. *Österreichische Ingenieur- und Architekten-Zeitschrift (ÖIAZ)*, 137(3).

van Impe, W.F. et al. (1994) Recent experiences and developments of the resonant vibrocompaction technique. In *XIII ICSMFE*, New Delhi, India.

Varaksin, S. (1990) Neuere Entwicklungen von Bodenverbesserungsverfahren und ihre Anwendung. In *5.Chr. Veder Kolloquium*, TU Graz, Styria, Austria.

Vesic, A.S. (1965) Ultimate loads and settlements of deep foundations in sand. In *Proceedings of the Symposium on Bearing Capacity and Settlement of Foundations in Sand*, Duke University, Durham, NC.

Vesic, A.S. (1972) Expansion of cavities in infinite soil mass. *JSMFD*. ASCE, 98.

Vrettos, C. (2009) Erschütterungsschutz. In Witt, K.J. (ed), *Grundbau-Taschenbuch, Teil 3, 7. Aufl.*, Ernst & Sohn, Berlin, Germany.

Vrettos, C. and Savidis, S. (2004) Seismic design of the foundation of an immersed tube tunnel in liquefiable soil. *Rivista Italiana di Geotechnica*, 38(4), 41–50.

Vucetic, M. (1994) Cyclic threshold shear strains in soils. *Journal of Geotechnical Engineering*, 120(12), 2208–2228.

Watts, K.S. et al. (2000) An instrumented trial of vibro ground treatment supporting strip foundations in variable fill. *Geotechnique*, 50(6), 699–708.

Weber, T.M. (2007) Modellierung der Baugrundverbesserung mit Schottersäulen. *Dissertationen der ETH Zürich*, Nr. 17321, Zürich, Switzerland.

Wehr, W. (1999) Schottersäulen—das Verhalten von einzelnen Säulen und Säulengruppen, *Geotechnik*, 22.

Wehr, W. (2005a) Influence of the carbonate content of sand on vibro compaction. In *6th International Conference on Ground Improvement Techniques*, Coimbra, Portugal.

Wehr, W. (2005b) Variation der Frequenz von Tiefenrüttlern zur Optimierung der Rütteldruckverdichtung. In *1. Hans Lorenz Symposium*, Heft 38, Grundbauinstitut der TU Berlin, Berlin, Germany.

Wehr, W. (2007) Rütteldruckverdichtung von karbonathaltigen Sanden. *Geotechnik-Kolloquium Freiberg*, Deutschland.

Wehr, J. and Sondermann, W. (2013) Deep vibro techniques. In Kirsch, K. and Bell, A. (eds.) *Ground Improvement*. CRC Press, Taylor & Francis Group, Boca Raton, FL.

Wehr, W. et al. (2008) Stone columns in very soft clays in Sweden. In *11th Baltic Sea Geotechnical Conference "Geotechnics in Maritime Engineering,"* Gdansk, Poland.

Weiss, K. (1978) 50 Jahre Deutsche Forschungsgesellschaft für Bodenmechanik (Degebo) 1928–1978. In *Mitteilungen der Deutschen Forschungsgesellschaft für Bodenmechanik an der Technischen Universität Berlin*, Heft 33, Berlin, Germany.

West, J.M. (1976) *The Role of Ground Improvement in Foundation Engineering: Ground Treatment by Deep Compaction*. The Institution of Civil Engineers, London, UK.

White, D. et al. (2002) *Constitutive Equations for Aggregates Used in Geopier Foundation Construction*. Department of Civil Construction Engineering, Iowa State University, Ames, IA.

Yip, T.C.W. et al. (2015) Offshore stone columns in soft clay at Hong Kong boundary crossing facility. In *International Conference on Soft Soil Engineering* (*ICSGE2015*), Singapore.

Yoshimi, Y. (1980) Protection of structures from soil liquefaction hazards. *Geotechnical Engineering*, 11.

Index